Brain Control of the Reproductive System

Brain Control of the Reproductive System

Edited by
Akira Yokoyama

JAPAN SCIENTIFIC SOCIETIES PRESS Tokyo

CRC PRESS Boca Raton Ann Arbor London Tokyo

Supported in part by the Ministry of Education, Science and Culture under the Grant-in-Aid for Publication of Scientific Research Result.

Published jointly by
JAPAN SCIENTIFIC SOCIETIES PRESS
2-10 Hongo, 6-chome, Bunkyo-ku, Tokyo 113, Japan
ISBN 4-7622-5696-X

and

CRC PRESS
2000 Corporate Blvd., N.W., Boca Raton, FL 33431, U.S.A.
ISBN 0-8493-7754-4

CIP Data available from the Library of Congress

Distributed in all areas outside Japan and Asia between Pakistan and Korea by CRC Press.

Printed in Japan

Preface

The reproductive activity, *i.e.* the sexual differentiation, sexual behavior and gonadal function including lactation, is indispensable for all mammals to preserve and propagate their species. Although the reproductive process seems to be complicated, we could understand the process more simply by viewing it in terms of how biological informations are transmitted and integrated in a network of the system. Geoffrey W. Harris's prediction in the 1940's on the existence of gonadotropin-releasing factor in the hypothalamus broke the dawn of the reproductive neuroendocrinology and first puts the brain in the highest position in this network operating the reproductive system. Informations originating from both the internal and external environments are conveyed to and integrated in the brain and transformed to humoral signals such as gonadotropin-releasing hormone (GnRH) that regulates the reproductive activity properly and efficiently through secretion of gonadotropins from the pituitary gland. Hormones secreted by the gonads in response to gonadotropins stimulation feed back to the brain and pituitary to control their activity. Thus, the brain, pituitary and gonads make up a dynamic axis. In other words, the reproductive activity is an expression of higher brain functions. In this context, studies on the brain functions are crucial in understanding precisely how reproductive activities are regulated and how they are modulated by various environmetal changes.

Although there are many aspects that must be discussed in the brain control of reproduction, the present book could not cover all of them. We, therefore, concentrate the discussion on the following points. Chapters 1, 2 and 3 mainly deal with morphological studies on the central nervous system, with special reference to the plasticity of GnRH neurons and sexually dimorphic cell groups, and estrogen receptors. These articles provide a basic knowledge of the sexual differentiation, the feedback mechanism of gonadal hormones, and the perception and processing of the conspecific signals such as pheromones. In Chapters 5 to 8, authors discuss the physiological mechanisms regulating pulsatile release of gonadotropin and oxytocin, and review the female sexual behavior. These articles would give readers a fundamental knowledge on the brain integration of either the internal and external factors, and the generator of pulsatile hormone secretion which is modulated by these factors.

Contributors of these topics are leading scientists in Japan, whose acquaintance I had made during my research career in Nagoya University from 1952 to 1990. They have been working actively on the brain function regulating the reproductive activity in mammals for many years and have published many outstanding papers in their field. They belong to institutions in various research fields, *i.e.* the medical sciences, veterinary medicine and animal sciences; and use a wide variety of approaches to investigate the integrating activity of the brain controlling reproduction. I believe, therefore, this book will be valuable for providing an up-to-date information of brain control of the reproductive system to graduate students and research workers interested in mammalian reproductive physiology. Furthermore, authors of this book inspires us to look at future research in areas of the brain control of the reproductive system which still remains unsolved.

Contemporary with Dr. Geoffrey W. Harris, Prof. Takashi Kobayashi worked on the neuroendocrine control of the reproductive system in mammals at Department of Obstetrics and Gynecology, Medical School, Tokyo Imperial University (now The University of Tokyo), and had published many outstanding papers suggesting the importance of the hypothalamus in regulating the reproductive activity. Dr. Kobayashi's works were published in Japanese, because papers were unable to publish in English at that time in Japan. Therefore, his contribution has been buried in oblivion. Armed with electrophysiological methods, the late Prof. Masazumi Kawakami, Department of Physiology, Medical School, Yokohama City University, and his group also made significant contributions in the brain function regulating the gonadal activity until his untimely death in 1982. I would like to mention their names here for commemorating their contribution to the neuroendocrine control of reproduction.

The help given by Mr. H. Matsuno and Mr. T. Miyazaki, Japan Scientific Societies Press, Tokyo, in the preparation of the book is most gratefully acknowledged.

Akira Yokoyama

Contents

1   Gonadotropin-Releasing Hormone Neurons in the Olfactory System

Masumi ICHIKAWA[1] **and Yoshitaka OKA**[2]

[1]*Department of Anatomy and Embryology, Tokyo Metropolitan Institute for Neuroscience, Fuchu, Tokyo 183*
[2]*Zoological Institute, Faculty of Science, The University of Tokyo, Tokyo 113*

Neuropeptides are known not only to control endocrine functions but also to modify the neuronal regulation of various kinds of behaviors such as sexual behavior (Renaud *et al.*, 1984; Meyerson, 1988). The decapeptide gonadotropin-releasing hormone (GnRH) may be one of the most extensively studied neuropeptides. In mammals, GnRH is involved in the regulation of luteinizing hormone (LH) and follicle stimulating hormone (FSH) secretion and probably in facilitation of the reproductive behavior. Several studies have shown that GnRH directly acts on certain parts of the brain to facilitate some components of reproductive behavior; *e.g.*, microinfusion of GnRH into the midbrain central grey facilitates the lordosis behavior of female rats (Sakuma & Pfaff, 1980). However, much yet remains to be done to clarify the neural mechanisms of such behavioral facilitation. Recently, there has been an increased interest in the olfactory system, since it contains GnRH neurons. In the present article, we describe the GnRH neuron in the olfactory system and introduce our study on the plasticity of GnRH fibers in the medial amygdaloid nucleus in rats (Ichikawa & Oka, 1989).

MORPHOLOGY AND DISTRIBUTION OF GNRH NEURONS

The GnRH neuron has a fusiform or bipolar-shaped cell body (Fig. 1). The neurons have been morphologically divided into two types in the rat brain (Jennes *et al.*, 1985; Takahashi *et al.*, 1988; Witkin, 1989; Witkin & Demasio, 1990): smooth GnRH cells are bipolar and have smoothly delineated cell bodies and dendritic shafts (Fig. 1A). Thorny or spiny GnRH cells are unipolar or bipolar and have irregularly-shaped cell bodies and dendrites (Fig. 1B, C). Witkin & Demasio (1990) reported that the

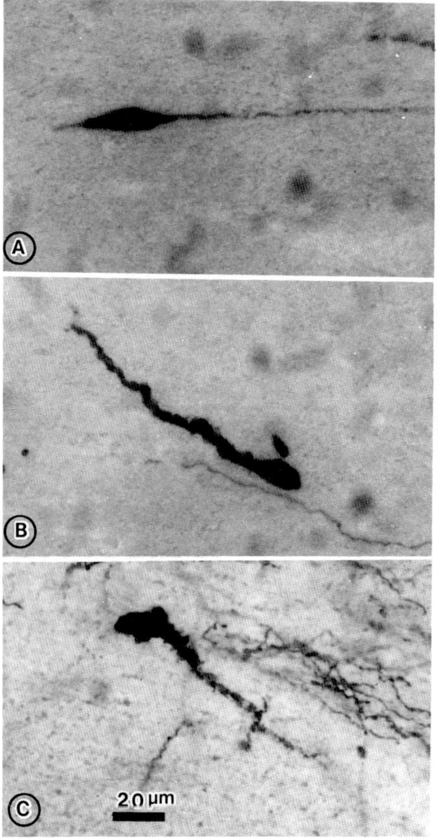

Fig. 1.   GnRH neurons in rat preoptic area. GnRH neurons were stained immunocytochemically with anti-GnRH antibody (LRH 13). (A) Smooth GnRH neuron. (B), (C) Thorny GnRH neurons.

smooth GnRH neurons contained less cytoplasm than the thorny neurons, but the size of their nuclei was the same. There were more and larger nucleoli in the smooth neurons. The thorny neurons contained more Golgi apparatus and more mitochondria, but the amount of endoplasmic reticulum was the same in the two types. On the basis of this morphology, Witkin & Demasio (1990) suggested that the smooth GnRH neurons might be more actively transcribing a message while the thorny GnRH neurons are more actively engaged in peptide processing and packing. In the goat brain, Hamada *et al.* (1991) reported multipolar cells as a third type of GnRH neuron. The functional difference between neuronal types has not yet been clarified. The fibers of GnRH neurons show beaded or varicose appearance and often a coiling style.

Distribution of GnRH cells and fibers has been studied intensively in rodents (Jennes & Stumpf, 1980; Witkin *et al.*, 1982; Merchenthaler *et al.*, 1984; Barry *et al.*, 1985; Silverman, 1984,1988). The GnRH cell bodies are mainly distributed in the septum, the diagonal band of Broca, the bed nucleus of stria terminalis, and the diencephalic area including the periventricular, medial preoptic, and anterior hypothalamic areas (Figs. 3A, 3B, 4). The greatest concentration of cell bodies is found between the ventral limb of the diagonal band of Broca at the level of the organum vasculosum of the lamina terminalis and the medial preoptic area at the level just posterior to the lamina terminalis. The projections of these cells have been studied immunocytochemically (Jennes & Stumpf, 1980; Witkin *et al.*, 1982; Merchenthaler *et al.*, 1984; Barry *et al.*, 1985; Silverman, 1984, 1988). The main projection is to the median eminence, which is the final common pathway for regulation of the anterior pituitary function. GnRH fibers reach the median eminence through more than one route. The major "septo-preoptico-infundibular" pathway is common to most species. GnRH cells in this area also project to the other hypothalamic areas and extra-hypothalamic sites. Some of the fibers which project to the median eminence leave the main pathway, enter the mamillary complex, converge on the mamillary peduncle, and proceed into the ventral tegmental area of the midbrain. Some GnRH cells in the septum and the preoptic area send fibers to the epithalamic region (habenula), continue into the fasciculus retroflexus, and terminate in the midbrain region. The cells in the bed nucleus of stria terminalis and the preoptic area send fibers to the medial amygdala through the stria terminalis.

The majority of recent studies on synaptology of GnRH neurons have used double-labeling procedures for GnRH and a transmitter (or its synthetic enzyme). Findings revealed a possible association between GnRH neurons and serotonin, dopamine, tyrosin hydroxylase, *etc.* (Palkovits *et al.*, 1982; Jennes *et al.*, 1983; Kiss & Halasz, 1985). Synaptic contacts of GnRH fibers with both GnRH cell bodies and non-GnRH neurons have been demonstrated in the preoptic area (Leranth *et al.*, 1985; Silverman & Witkin, 1985). These synapses were characterized by a well defined synaptic cleft, as well as an accumulation of both large immunoreactive granules and small electron lucent non-immunoreactive vesicles. In the rat midbrain central grey, Buma (1989) observed GnRH fibers by electron microscopy but found no GnRH synapses. He suggested that the GnRH was released non-synaptically.

As described in this section, GnRH fibers which do not terminate in the median eminence have been observed in the central nervous system. Such "extrahypothalamic" GnRH fibers have been observed in every vertebrate species examined to date. The most clear dichotomy of the preoptic-hypophysial and other GnRH systems has been demonstrated in a fish, the dwarf gourami (Oka & Ichikawa, 1990). In this animal, GnRH neurons are distributed in two cell clusters, one in the transitional area between the olfactory bulb and the telencephalon (ganglion cells of the terminal nerve, TN cells), and the other in the preoptic area. After making bilateral electrolytic lesions of the TN cells, most of the GnRH immuno-reactive (GnRH-IR) fibers disappeared, whereas GnRH-IR cell bodies and axons comprising the preoptic-hypophysial system remained intact (Yamamoto, Oka & Kawashima, unpublished observations). Thus, the TN cells are a major component of the GnRH system, at least in the fish brain, and the TN-GnRH system is structurally independent from the preoptic/hypophysial-GnRH system which facilitates gonadotropin release from the pituitary. We have therefore hypothesized that GnRH released from TN cells functions as an important neuromodulator.

It has been reported that GnRH evokes a late, slow EPSP in the bullfrog sympathetic ganglion cells by inhibiting one type of potassium current, the M-current (Jones & Adams, 1987). The inhibition of the M-current increases the excitability of the cell and allows repetitive firing in response to a maintained depolarizing current. A similar phenomenon (the reduction of the spike frequency adaptation caused by GnRH) has been reported in rat hippocampal pyramidal cells (Wong *et al.*, 1990). It has also been reported in the rat that GnRH potentiates responses of preoptic neurons to norepinephrine and serotonin (Pan *et al.*, 1988). Thus, GnRH seems to act as a neuromodulator in the peripheral as well as the central nervous system to increase the excitability of the cell. However, the biological significance of this GnRH action has not been clarified. Sakuma & Pfaff (1980) reported that the infusion of GnRH into the midbrain central grey can potentiate the lordosis behavior of female rats. Shivers *et*

al. (1983) demonstrated that the gonadal steroid affects the contents of
GnRH in the GnRH fibers of the midbrain central grey. These findings
suggest that these fibers distributed in the central grey are related to the
reproductive function. Therefore, more extensive studies are necessary, first
on the behavioral and physiological effects of the GnRH neuronal system
and then on the cellular and molecular mechanisms of such biological
effects.

GNRH NEURONS IN THE OLFACTORY SYSTEM

The olfactory system in mammals has two components (Scalia & Winans,
1975): the main and accessory olfactory systems (Fig. 2). The receptor
neurons for the main olfactory system are distributed throughout the
olfactory epithelium and project into the main olfactory bulb (MOB),
while the receptor neurons for the accessory olfactory system lie in the
vomeronasal organ (VNO). The receptor neurons in the VNO project to the
accessory olfactory bulb (AOB). The main projection neurons in the MOB

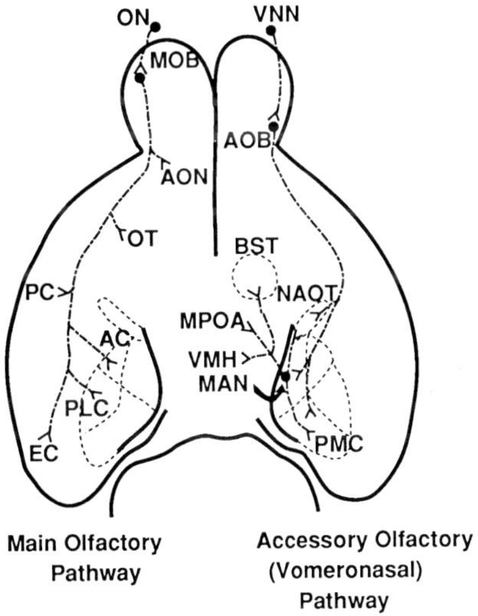

Fig. 2. Schematic representation of the olfactory pathway. Left, main olfactory pathway; right, accessory olfactory pathway. AC, anterocortical amygdaloid nucleus; AOB,
accessory olfactory bulb; AON, anterior olfactory nucleus; BST, bed nucleus of stria
terminalis; EC, entorhinal cortex; MAN, medial amygdaloid nucleus; MOB, main
olfactory bulb; MPOA, medial preoptic area; NAOT, bed nucleus of accessory olfactory
tract; ON, olfactory nerve; PC, prepiriform cortex; PMC, posteromedial cortical amygdaloid nucleus; VMH, ventromedial hypothalamus; VNN, vomeronasal nerve.

are the mitral cells. The receptor cells in the olfactory epithelium terminate
on the dendrites of mitral cells in the glomeruli of the MOB. The efferent
pathway of the MOB is the lateral olfactory tract, and the terminal fields are

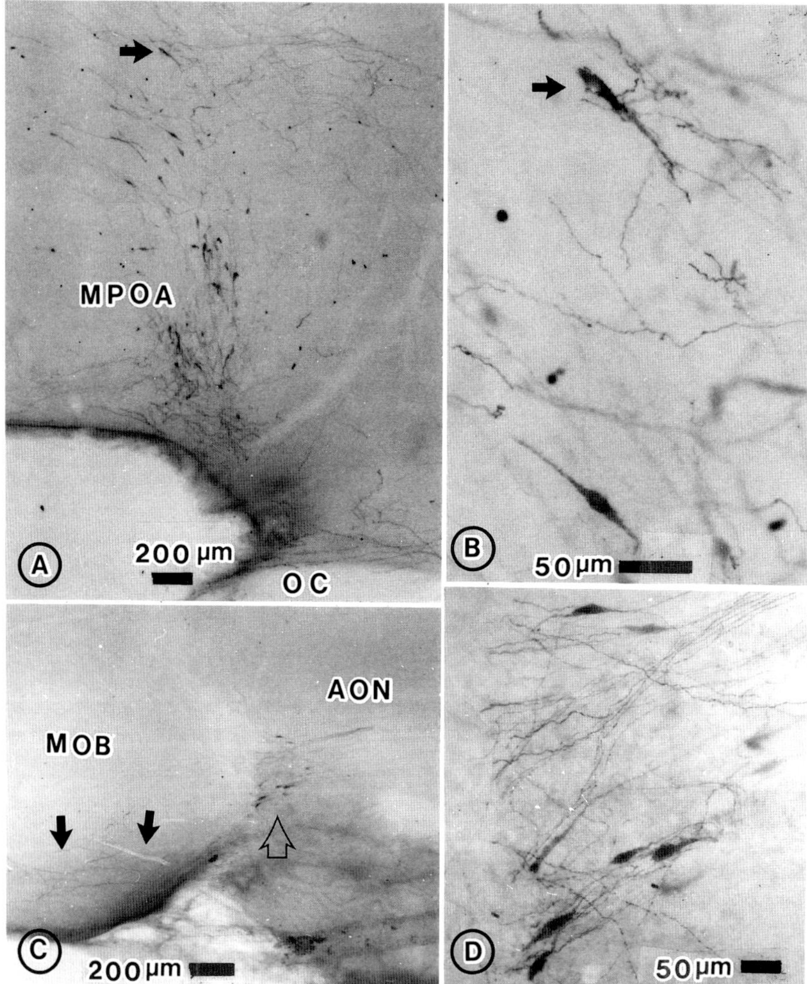

Fig. 3. GnRH neurons stained immunocytochemically using anti-GnRH antibody
(LRH 13). (A) GnRH neurons in the medial preoptic area (MPOA). (B) Higher
magnification of (A). (C) Photograph showing the ventral region of transitional area
between the main olfactory bulb (MOB) and the anteror olfactory nucleus (AON). Filled
arrow indicates the GnRH neurons in the ganglion of the terminal nerve (GTN). Open
arrow shows the GnRH fibers in the ventral region of MOB. (D) GnRH neurons in the
GTN of Fig. C.

the anterior olfactory nucleus, olfactory tubercles, prepiriform cortex, anterior and posterolateral cortical amygdaloid nuclei, and entorhinal cortex. The projection neurons in the AOB receive the projection from the vomeronasal receptor neurons and terminate in the medial and postero-medial amygdaloid nuclei. It has been reported that the interruption of the accessory olfactory pathway results in the alteration and elimination of the expression of various behaviors mediated by the conspecific chemical cues (pheromone). Therefore, the accessory olfactory system has been thought to play an important role in perception and processing of conspecific chemical signals in mammals (Wysocki, 1979; Keverne, 1983; Meredith, 1983; Halpern, 1987).

The GnRH cells and fibers are observed in both main and accessory olfactory systems in rodents (Phillips *et al.*, 1980, 1982; Witkin & Silverman, 1983; Jennes, 1986, 1987; Ichikawa & Oka, 1989). The morphological characteristics of GnRH neurons in this system are almost the same as those

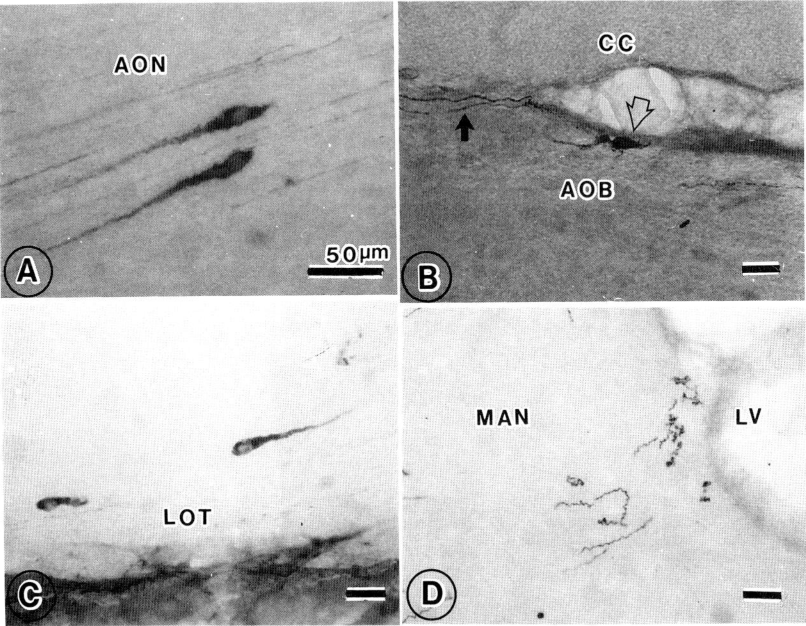

Fig. 4. Immunocytochemically stained GnRH neurons. (A) Bipolar GnRH neurons observed in the anterior olfactory nucleus (AON). (B) GnRH cell body observed in the accessory olfactory bulb (AOB) (open arrow). Filled arrow indicates the GnRH fibers in AOB. CC, cerebral cortex. (C) GnRH cell bodies observed in the lateral olfactory tract (LOT). (D) GnRH fibers observed in the medial amygdaloid nucleus (MAN). LV, lateral ventricle.

in the septum, the preoptic area and the hypothalamus. A group of GnRH cell bodies is found in the ventral surface of the transitional area between the olfactory bulb and the anterior olfactory nucleus (Fig. 3C and D). This group is called the ganglion of the terminal nerve (Schwanzel-Fukuda & Silverman, 1980). Some GnRH cells are also observed along the accessory olfactory pathway: the vomeronasal nerve, the accessory olfactory bulb

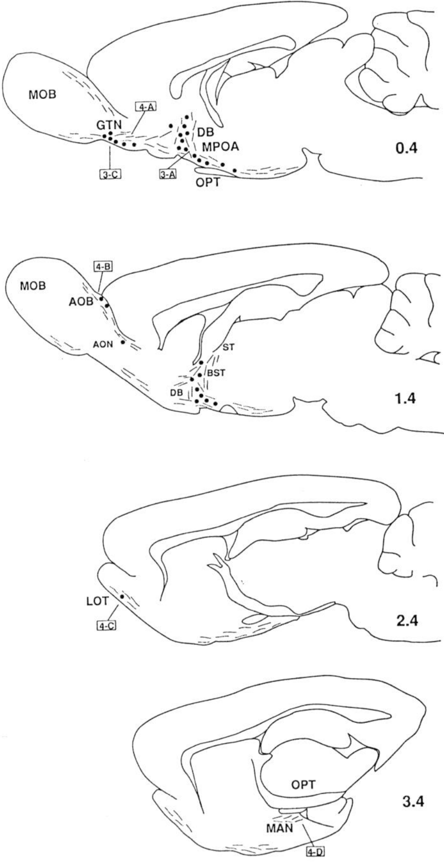

Fig. 5. Schematic drawing of sagittal sections of rat brain showing the distribution of the GnRH cell bodies (filled circles) and the GnRH fibers (broken lines). Numbers and letters in squares indicate the regions where each figure in this article was observed. Numbers at right show the distance (mm) from the midline. AOB, accessory olfactory bulb; AON, anterior olfactory nucleus; BST, bed nucleus of stria terminalis; DB, diagonal band of Broca; GTN, ganglion of terminal nerve; LOT, lateral olfactory tract, MAN, medial amygdaloid nucleus; MOB, main olfactory bulb; MPOA, medial preoptic area; OPT, optic tract.

(Fig. 4B), and the accessory olfactory tract. A small number of GnRH cells are observed in the lateral olfactory tract (Fig. 4C) and the anterior olfactory nucleus (Fig. 3A); no GnRH cell body is observed in the other areas of the main olfactory pathway. GnRH fibers are distributed widely in both olfactory pathways (Fig. 4). It is difficult to trace the course of each GnRH fiber and to identify the fibers' original cells. Synaptic terminals have been reported in the AOB (Philips *et al.*, 1982). The distribution of GnRH cells and fibers in the rat olfactory system is drawn diagramatically in Fig. 5. The close anatomical relation of the GnRH neurons to the olfactory system suggests that the GnRH is involved in the modulation of processing of the olfactory information. Especially, the relationship between the GnRH and the accessory olfactory pathway implies that the accessory olfactory system has an important role in the regulation of reproductive function.

PLASTICITY OF GNRH FIBERS IN THE MEDIAL AMYGDALOID NUCLEUS

The accessory olfactory pathway starts from the vomeronasal organ (VNO), passes through the AOB, reaches the medial amygdala (medial amygdaloid nucleus (MAN) and posteromedial cortical amygdaloid nucleus, *etc*), and projects to the preoptic area and/or ventromedial hypothalamus (Fig. 2). We previously demonstrated a lesion-induced reorganization of neuronal connections in the MAN following denervation of the afferent fibers from the AOB in adult rats (Ichikawa, 1987a, b, 1988). The formation of synapses by the residual fibers (fibers from the bed nucleus of the stria terminalis and posteromedial cortical amygdaloid nucleus) takes place in the MAN after denervation of the AOB fibers. We have also demonstrated a recovery of the olfactory behavior following removal of the AOB in adult rats. There, we suggested the possibility that the reorganization of neuronal connections in the MAN was related to the recovery of the behavior after the AOB removal (Ichikawa 1989). The mechanism of functional recovery after the injury still remains unknown (Steward, 1982; Goldbeger & Murray, 1985; Marshall, 1985).

It has been demonstrated that the GnRH fibers are distributed in the MAN (Phillips *et al.*, 1980; Witkin & Silverman, 1983; Rosser *et al.*, 1986; Jennes, 1986; Ichikawa & Oka, 1988; Dudley *et al.*, 1990), though their role has not yet been clarified. We examined whether the GnRH fibers in MAN undergo plastic changes after a denervation of the AOB fibers.

The AOB of 23 adult rats was removed unilaterally by suction to denervate the AOB fibers from the MAN, and the unilateral olfactory bulb (MOB+AOB) of 15 rats was removed in the same way. In 10 rats, only the dura mater was cut with scissors, but the brain parenchyma was not damaged (sham operation). The contralateral olfactory bulb of all operated rats was kept intact as a control in each operated animal. Nine rats were used for the intact control group. In the AOB removal, the operated rats were grouped into three according to survival time: the survival time of animals in the first group was 6 hours (7 rats), 4 days in the second one (8 rats) and 2 months in the third one (8 rats). In the AOB plus MOB removal, the operated rats were in two groups: the 4 day survival time group (8 rats) and the 2 month group (7 rats). In the sham operated group, all rats survived for 4 days. At autopsy, all animals were anesthetized with Nembutal and perfused with saline followed by Zamboni's fixative. The

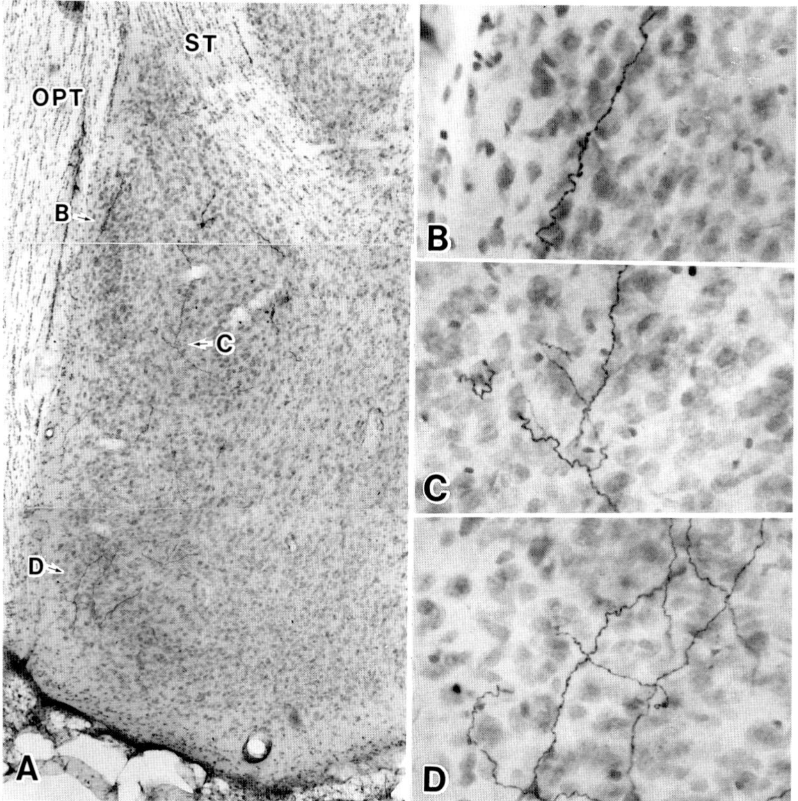

Fig. 6.   GnRH-immunoreactive fibers in the medial amygdaloid nucleus. Letters B, C, and D in A indicate the region where the fibers in B, C, and D were observed. Beaded swellings and coilings of fibers are evident in B, C, and D. OPT: optic tract, ST: stria terminalis. Magnification (A): ×65, (B), (C), (D): ×260.

fixed brains were cut serially with a freezing microtome and frontal sections including the MAN were selected and processed for immunocytochemical staining using an antibody for GnRH (LRH 13) (Oka & Ichikawa, 1990). One section was selected from 3 serial sections including the posterior half of the MAN. Four sections were chosen from each animal. GnRH-immunoreactive (GnRH-IR) fibers in the MAN of each section were traced with the aid of a camera lucida. The length of the fibers traced was measured using a digitizer connected to a personal computer. Length of the GnRH-IR fibers in 4 sections was totalled in each animal; this cumulative length is tentatively called "total" length of GnRH-IR fibers. These values were compared between the operated and control sides and also between each operated group. Statistical analysis of difference between the operated and control sides was done in each operated group using the Wilcoxon matched paired test. The ratio of the length in the operated and control sides was calculated in each animal, and the difference between each

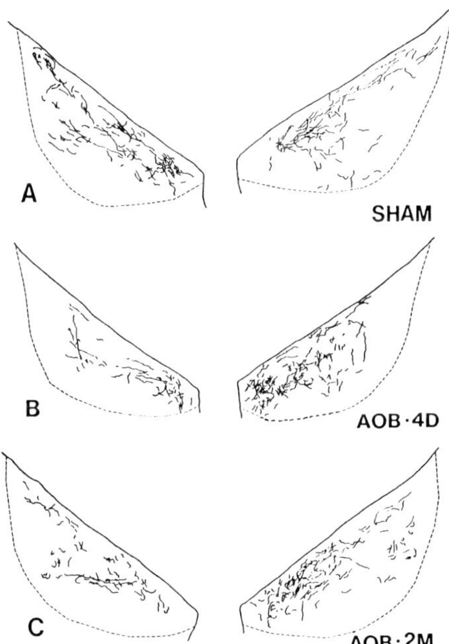

Fig. 7. Camera lucida drawings of GnRH-IR fibers in the medial amygdaloid nucleus. A: sham-operated rats. B: 4 day survivors after AOB removal. C: 2 month survivors after AOB removal. The operated side is at right and the intact side at left. Tracings of four sections were superimposed on one drawing.

experimental group and the intact and/or sham groups was statistically analyzed by Mann-Whitney U test.

In all rats examined, the GnRH-IR fibers were distributed in the MAN (Fig. 6). Examples of the distribution in some groups are shown in Fig. 7.

Total length of GnRH-IR fibers in the MAN of each rat is shown in Fig. 8. In 5 intact rats, the length of the fibers on the left side was longer than that on the right side, and in the other 4 intact rats the length on the right side was longer than that on the left. Thus, the difference in length of GnRH-IR fibers between the left and right sides was not statistically significant in the intact group. In the sham-operated group, the length of fibers of the sham-operated side in 6 rats was longer than that of the control side, but in the other 4 sham-operated rats the result was the reverse. The length of GnRH-IR fibers on the operated side was thus not different statistically from that on the intact side in sham-operated rats.

At 6 hours after the AOB removal, the length of GnRH-IR fibers on the operated side was not different from that on the intact side (the operated side was longer than the intact one in 4 rats, but the operated side was

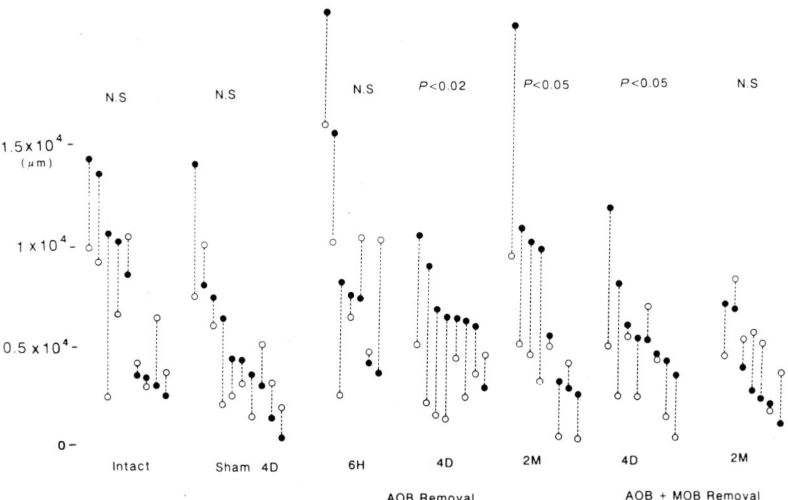

Fig. 8. Total length of GnRH fibers observed in the medial amygdaloid nucleus in each rat. Filled circles show the length of the fibers on the operated side, and open circles that on the intact side. The rats are divided into 7 groups according to the kind of operation and the survival time (intact, 4 days survival after sham operation (4D-sham), 6 hours (6H), 4 days (4D), and 2 months (2M) survival after AOB removal, 4 days and 2 months survival after AOB+MOB removal. Ordinate indicates the length of GnRH-IR fibers. Statistic difference between operated and intact sides was tested in each group using Wilcoxon matched pairs test.

shorter than the intact one in 3 rats). Four days after the AOB removal, in 7 rats out of the 8 surviving, the length of GnRH-IR fibers on the operated side was longer than that on the intact side. Only one rat showed the reverse result. In fact, the length of GnRH-IR fibers on the operated side was statistically different from that on the intact side 4 days after the AOB removal ($P < 0.02$). At 2 months survival time, in 7 rats out of 8, the length of GnRH-IR fibers on the operated side was longer than that on the intact side. The difference between the operated side and the intact side was statistically significant ($P < 0.05$).

In 7 of 8 rats surviving for 4 days after the AOB + MOB removal, the length of GnRH-IR fibers on the operated side was longer than that on the intact side. In only one rat, the length of GnRH-IR fibers on the operated side was shorter than that on intact side. The difference of GnRH-IR fibers between the operated and intact side was significant ($P < 0.05$). At 2 months survival time, in 5 rats out of 7, the length of GnRH-IR fibers on the intact side was longer than that on the operated side. The length of GnRH-IR fibers on the two sides was not statistically different.

Figure 9 shows the ratio of the length of GnRH-IR fibers on the operated side to that on the intact side in each experimental group. In the

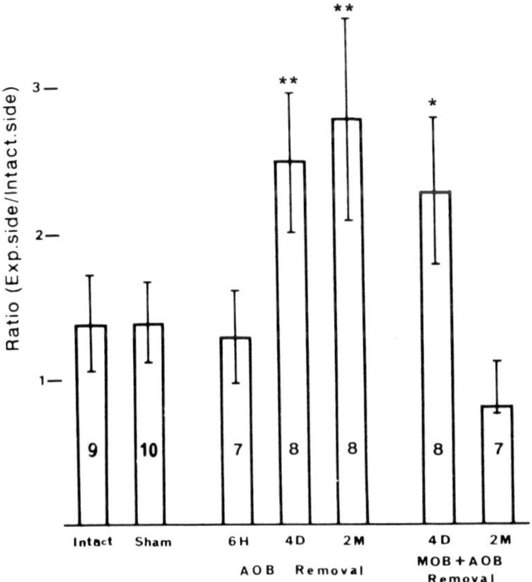

Fig. 9. Ratio of the length of GnRH-IR fibers on the operated side to that on the intact side. The ratio at 4 days and 2 months survival after AOB removal and 4 days survival after AOB + MOB removal is significantly higher than that of the sham-operated groups (Mann-Whitney U test). *$P < 0.1$, **$P < 0.05$.

intact group, the ratio of fiber-length on the left side to that on the right was 1.38 ± 0.35. The ratio in the sham-operated group was 1.39 ± 0.28 which was not different from that of the intact group. In the AOB removal group, the ratio was 1.29 ± 0.32 at 6 hours survival time, 2.49 ± 0.47 at 4 days survival time, and 2.79 ± 0.70 at 2 months survival time. The values at both 4 days and 2 months survival time were statistically different from that of the sham-operated group ($P < 0.05$). In the AOB plus MOB removal group, the ratio was 2.29 ± 0.52 at 4 days survival time and 0.81 ± 0.18 at 2 months survival time. The value at 4 days survival time showed that the GnRH-IR fibers in the operated side had a tendency to be longer than those in the intact side. But the value at 4 days and 2 months survival groups was not different significantly from that of the sham-operated one.

These results show that the total length of GnRH-IR fibers on the operated side is longer than that on the intact side at 4 days and 2 months after the AOB removal.

We earlier demonstrated the reorganization of neuronal connections in the MAN of adult rats (Ichikawa, 1987a, b, 1988). Formation of new neuronal connections by residual fibers (fibers from the stria terminalis and fibers from the posteromedial cortical amygdaloid nucleus) takes place in the MAN following the denervation of AOB fibers. The present study shows, moreover, the increase in length of GnRH-IR fibers in the MAN after the AOB removal. There are some possible explanations for the increase in GnRH-IR fibers following the denervation of AOB fibers. (1) The denervation of AOB fibers could affect the GnRH neurons. The synthesis and/or transportation of GnRH is somehow enhanced in the GnRH neurons after the denervation, and then the GnRH content increases in the fibers in which GnRH could not be detected immunocytochemically under normal condition. This change may be transient. Consequently, the "GnRH-IR" fibers apparently increase in the MAN but the real number of fibers does not change (Fig.10B). (2) The release of GnRH may be impaired in the MAN after the denervation of the AOB fibers, while the GnRH synthesis remains unchanged. In this case, the content of GnRH in the fibers of the MAN increases. Thus, the GnRH-IR fibers appear to increase, but the real number of fibers does not change in the MAN (Fig. 10B). (3) The sprouting of the GnRH fibers takes place in the MAN after denervation of the AOB fibers in a similar manner as the other remaining afferent fibers such as the fibers from the bed nucleus of the stria terminalis (Ichikawa, 1987b) and the posteromedial corticoamygdaloid nucleus (Ichikawa,1988). Consequently the number of GnRH-IR fibers actually increases in the MAN (Fig.10C). In the present study, however, it could not be clarified which event takes place in the MAN.

At 4 days survival time, in both the AOB and the AOB+MOB removed rats, the number of these fibers increased in the MAN. The increase in length of GnRH-IR fibers at 4 days survival time may be a short-term and transient phenomenon. We suggest that this short-term increase takes place because of either (1) the transient enhancement of the synthesis and/or transportation of GnRH or (2) the transient inhibition of GnRH release from fibers (Fig.10B). The increase of GnRH-IR fibers

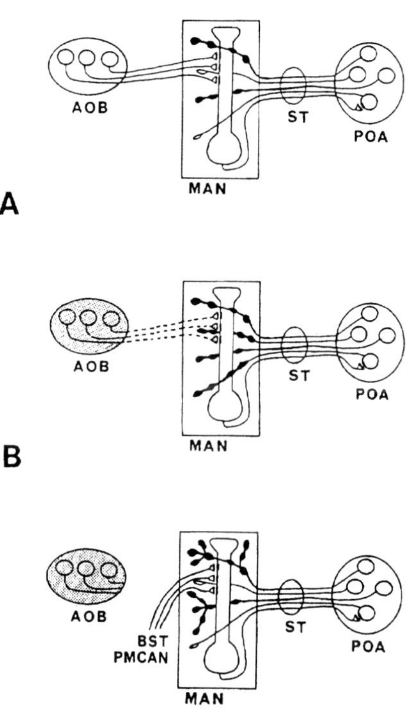

Fig. 10.   Hypothetical drawing of plasticity of GnRH fibers in the medial amygdaloid nucleus (MAN) after the removal of the accessory olfactory bulb (AOB). Thick and beaded fibers indicate the GnRH-IR fibers. (B) Short-term change of GnRH fibers in MAN after AOB removal. The content of GnRH-IR increase in fiber terminals. Shaded area indicates the removed area, and broken lines show the degenerating fibers. (C) Long-lasting change of GnRH-IR fibers in MAN after AOB removal. The terminals of GnRH-IR fibers sprout. The fibers from bed nucleus of stria terminalis (BST) and the posteromedial cortical amygdaloid nucleus (PMCAN) occupy the denervated sites (Ichikawa, 1987b, 1988, 1990). POA, preoptic area; ST, stria terminalis.

continued for 2 months after the AOB removal, and this long-lasting increase of GnRH fibers could be due to the sprouting of fibers (Fig. 10C). Recently, Phelps & Saporta (1988) reported that the sprouting of GnRH fibers occurs in the hypothalamus after injury. Gotow *et al.*(1989) demonstrated that somatostatin-immunoreactive fibers sprout and form new synapses in the central nucleus of the amygdala in rats after partial deafferentation. Thus, it is possible that the sprouting of GnRH fibers takes place in the MAN after the denervation of AOB fibers.

It has been suggested that the accessory olfactory pathway is involved in the perception and processing of conspecific chemical signals (pheromonal information), and that the MAN has an important role in the regulation of reproduction, especially the reproductive behavior mediated by the intersexual odor (Wysocki, 1979; Keverne, 1983; Kostarczyk, 1986; Halpern, 1987). We have demonstrated that the preference behavior of the intersexual odor is impaired just after the AOB removal, gradually recovers, and reaches about 80% of the intact level at least 1 month after the removal (Ichikawa, 1989). It was suggested that the reorganization of neuronal connections in the MAN following AOB removal is one of the important events mediating the recovery of behavior. On the other hand, the AOB+ MOB removed rats show no recovery at any survival time. In the present study, at 2 months survival time, the number of GnRH-IR fibers increased in the MAN of AOB removed rats, but not in the case of the AOB+MOB removal. This suggests that the long-lasting increase of GnRH-IR fibers could mediates the functional recovery in the accessory olfactory system.

The plastic change of GnRH-IR fibers seems to be an important factor in recovery after injury. Further studies, however, are necessary to elucidate the relationship between the increase of GnRH fibers and the functional recovery.

CONCLUSION

The GnRH cell bodies and fibers are densely distributed in the olfactory system. A larger number of GnRH neurons are found in the accessory olfactory system than in the main olfactory system. It has been argued that the accessory olfactory system is involved in perception and processing of pheromonal signals. Thus, it is suggested that the GnRH neurons play a role in the regulation of olfactory-related reproductive function. We reported here the plastic changes (sprouting) of GnRH fibers in the medial amygdaloid nucleus after elimination of the afferent fibers from the AOB and suggested the relationship between the plasticity and functional recovery. In the next step, it is necessary to examine the function of GnRH neurons in the olfactory system using electrophysiological, pharmacological, and biochemical techniques as well as determination of the relationship between the plasticity of GnRH fibers and the functional recovery.

ACKNOWLEDGEMENT

The authors would like to thank Dr. K. Wakabayashi, Gunma University and Dr. M.K. Park, University of Tokyo for the supply of a monoclonal antibody (LRH 13).

REFERENCES

Barry, J., Hoffman, G.E, & Wray, S. (1985). LHRH-containing systems. In: *Handbook of Chemical Neuroanatomy*, Vol. 4: GABA and Neuropeptides in the CNS, Part I. (Bjorklund, A. and Hokfelt, T., eds.), Elsevier, Amsterdam, pp. 166–215.

Buma, P. (1989). Characterization of luteinizing hormone-releasing hormone fibres in the mesencephalic central grey substance of the rat. *Neuroendocrinology* **49**, 623–630.

Dudley, C.A., Lee Y. & Moss, R.L. (1990). Electrophysiological identification of a pathway from the septal area to medial amygdala: sensitivity to estrogen and LHRH. *Synapse* **6**, 161–168.

Goldbeger, M.E. & Murray, M. (1985). Recovery of function and anatomical plasticity after damage to the adult and neonatal spinal cord. In: *Synaptic Plasticity* (Cotman, C.W., ed.), Guilford, New York, pp. 77–110.

Gotow, T., Williams, T.H., Jew, J.Y., Cassell, M.D., Palkovits, M. & Hashimoto, P.H. (1989). Collateral sprouting of somatostatin-immunoreactive axons after partial deafferentation of the central nucleus of the rat amygdala. *Brain Research* **492**, 325–336.

Halpern, M. (1987). The organization and function of the vomeronasal system. *Annual Review of Neuroscience* **10**, 325–362.

Hamada, T., Shimizu, T., Mori, Y. & Ichikawa, M. (1991). Immunohistochemical study on GnRH neurons in the Shiba goat. *Proceedings of the 15th Annual Meeting of Japan Society for Comparative Endocrinology.*

Ichikawa, M. (1987a). Synaptic reorganization in the medial amygdaloid nucleus after lesion of the accessory olfactory bulb of adult rat.I. Quantitative and electron microscopic study of the recovery of synaptic density. *Brain Research* **420**, 243–252.

Ichikawa, M. (1987b). Synaptic reorganization in the medial amygdaloid nucleus after lesion of the accessory olfactory bulb of adult rat. II. New synapse formation in the medial amygdaloid nucleus by fibers from the bed nucleus of the stria terminalis. *Brain Research* **420**, 253–258.

Ichikawa, M. (1988). Plasticity of intra-amygdaloid connections following the denervation of fibers from accessory olfactory bulb to medial amygdaloid nucleus in adult rat: immunohistochemical study of anterogradely transported lectin (*Phaseolus vulgaris* leucoagglutinin). *Brain Research* **451**, 248–254.

Ichikawa, M. (1989). Recovery of olfactory behavior following removal of accessory olfactory bulb in adult rat. *Brain Research* **498**, 45–52.

Ichikawa, M. (1990). Plasticity of neuronal connections in medial amygdaloid nucleus after removal of accessory olfactory bulb in adult rat. In: *Vision, Memory, and the Temporal Lobe* (Iwai, E., ed.), Elsevier, New York, pp. 399–403.

Ichikawa, M. & Oka, Y. (1988). LHRH immunoreactive neurons in the rat olfactory system. *Acta Anatomica Nipponica* **63**, 391.

Ichikawa, M. & Oka, Y. (1989). Plasticity of LHRH fibers in the olfactory pathway. *Neuroscience Abstract* **19**, 976.

Jennes, L. (1986). The olfactory gonadotropin releasing hormone immunoreactive system in mouse. *Brain Research* **386**, 351–363.

Jennes, L. (1987). Sites of origin of gonadotropin releasing hormone containing projections to the amygdala and interpeduncular nucleus. *Brain Research* **404**, 1339–1344.

Jennes, L. & Stumpf, W. E. (1980). LHRH-systems in the brain of the golden hamster. *Cell and Tissue Research* **209**, 239–256.

Jennes, L., Stumpf, W.E. & Trappaz, M.L. (1983). Anatomical relationships of dopaminergic and GABAergic system with the GnRH system in the septo-hypothalamic area. *Experimental Brain Reserch* **50**, 91–99.

Jennes, L., Stumpf, W.E. & Sheedy, M.E. (1985). Ultrastructural characterization of GnRH producing neurons. *Journal of Comparative Neurology* **232**, 534–547.

Jones, S.W., & Adams, P.R. (1987) In: *Neuromodulation. The Biochemical Control of Neuronal Excitability* (Kaczmarek, L.K. and Levitan, I.B., eds.), Oxford University Press, New York/Oxford, pp. 159–186.

Keverne, E.B. (1983). Pheromonal influences on the endocrine regulation of reproduction. *Trends in Neuroscience* **6**, 381–384.

Kiss, J. & Halasz, B. (1985). Demonstration of serotonergic axon termination on luteinizing hormone-releasing hormone in the preoptic area of rat using a combination of immunocytochemistry and high resolution ARG. *Neuroscience* **14**, 69–78.

Kostarczyk, E.M. (1986). The amygdala and male reproductive functions: I. Anatomical and endocrine bases. *Neuroscience and Biobehavioral Reviews* **10**, 67–77.

Leranth, C., Seguraum, L.M.G., Palkovits, M., MacLusky, N.T., Shanabraegh, M. & Naftolin, F. (1985). The LH-RH containing neuronal network in the preoptic area of the rat: Demonstration of LH-RH containing nerve terminals in synaptic contact with LH-RH neurons. *Brain Research* **345**, 332–336.

Marshall, J.F. (1985). Neural plasticity and recovery of function after brain injury. *International Review of Neurobiology* **26**, 201–247.

Merchenthaler, I., Gorcs, T., Setalo, G., Petrusz, P. & Flerko, B. (1984). Gonadotropin-releasing hormone (GnRH) neurons and pathways in the rat brain. *Cell and Tissue Research* **237**, 15–29.

Meredith, M. (1983). Sensory physiology of pheromone communication. In: *Pheromones and Reproduction in Mammals* (Vandenbergh, J.G., ed.), Academic Press, pp. 200–252.

Meyerson, B (1988). Neuropeptides and sexual function. In: *Handbook of Sexology*, (Sitsen, J.M.A., ed), Vol 6, Elsevier, New York, pp. 44–65.

Oka, Y. & Ichikawa, M. (1990). Gonadotropin-releasing hormone (GnRH) immunoreactive system in the brain of the dwarf gourami (*Colisa lalia*) as revealed by light microscopic immunocytochemistry using a monoclonal antibody to common amino acid sequence of GnRH. *Journal of Comparative Neurology* **300**, 511–522.

Palkovits, M., Leranth, C., Jew, J.Y. & Williams, T. (1982). Simultaneous characterization of pre- and postsynaptic neuron contact sites in brain. *Proceedings of National Academy of Science, U.S.A.* **79**, 2705–2708

Pan, J.T., Kow, L.M. & Pfaff, D.W. (1988). Modulatory actions of luteinizing hormone-relcasing hormone on electrical activity of preoptic neurons in brain slice. *Neuroscience* **27**, 623–628.

Phelps, C.P. & Saporta, S. (1988). The release of pituitary LH and sprouting of LH-releasing hormonc (LHRH)-containing neurons after anterior hypothalamic deafferentation. *Brain Research* **454**, 188–204.

Phillips, H.S., Hostetter, G., Kerdelhue, B. & Kozlowski, G.P. (1980). Immunocytochemical localization of LHRH in central olfactory pathways of hamster. *Brain Research* **193**, 574–579.

Phillips, H.S., Ho, B.T. & Linner, J.G. (1982). Ultrastructural localization of LH-RH immunoreactive synapses in hamster accessory olfactory bulb. *Brain Research* **246**, 143–204.

Renaud, L.P.,Day, T.A. & Ferguson, A.V. (1984). CNS regulation of reproduction: Peptidergic mechanisms. *Brain Research Bulletin* **12**, 181–186.

Rosser, A.E., Hokefelt, T. & Goldstein, M. (1986). LHRH and catecholamine neuronal systems in the olfactory bulb of the mouse. *Journal of Comparative Neurology* **250**, 352–363.

Sakuma, Y. & Pfaff, D. W. (1980). LH-RH in the mesencephalic central grey can potentiate lordosis reflex of female rats. *Nature* **283**, 566–567.

Scalia, F. & Winans, S.S. (1975). The differential projections of the olfactory bulb and accessory olfactory bulb in mammals. *Journal of Comparative Neurology* **161**, 31–56.

Schwanzel-Fukuda, M. & Silverman, A.J. (1980). The nervus terminalis of the guinea pig: a new luteinizing hormone-releasing hormone (LHRH) neuronal system. *Journal of Comparative Neurology* **191**, 213–225.

Shivers, B.D., Harlan, R.E., Morrell, J.I. & Pfaff, D.W. (1983). Immunocytochemical localization of luteinizing hormone-releasing hormone in male and female rat brains. *Neuroendocrinology* **36**, 1–12.

Silverman, A.J. (1984). Luteinizing hormone releasing hormone containing synapse in the diagonal band and preoptic area of guinea pig. *Journal of Comparative Neurology* **227**, 452–458.

Silverman, A.J. (1988). The gonadotropin-releasing hormone (GnRH) neuronal systems: Immunocytochemistry. In: *The Physiology of Reproduction* (Knobil, E. and Neil, J., eds.), Raven Press, New York, pp. 1283–1304.

Silverman, A.J. & Witkin, J.W. (1985). Synaptic interactions of LHRH neurons in the guinia pig preoptic area. *Journal of Histochemistry and Cytochemistry* **33**, 66–72.

Steward, O. (1982). Assessing the functional significance of lesion-induced neuronal plasticity. *International Review of Neurobiology* **23**, 197–254.

Takahashi, S., Ono, R., Nomura, K. & Kawashima, S. (1988). Luteinizing hormone-releasing hormone (LHRH) neurons in the male and female rats at peripubertal period. *Anatomy and Embryology* **178**, 475–480.

Witkin, J.W. (1989). Morphology of LHRH neurons as a function of age and hormonal condition in the male rat. *Neuroendocrinology* **49**, 344–348.

Witkin, J.W., Paden, C.M. & Silverman, A. J. (1982). The luteinizing hormone-releasing hormone (LHRH) systems in the rat brain. *Neuroendocrinology* **35**, 429–438.

Witkin, J.W. & Silverman, A.J. (1983). Luteinizing hormone releasing hormone (LHRH) in rat olfactory system. *Journal of Comparative Neurology* **218**, 426–432.

Witkin, J.W. & Demasio, K. (1990). Ultrastructural differences between smooth and thorny gonadotropin releasing hormone neurons. *Neuroscience* **34**, 777–783.

Wong, M., Eaton, M.J. & Moss, R.L. (1990). Electrophysiological actions of luteinizing hormone-releasing hormone: intracellular studies in the rat hippocampal slice preparation. *Synapse* **5**, 65–70.

Wysocki, C. J. (1979). Neurobehavioral evidence for the involvement of the vomeronasal system in mammalian reproduction. *Neuroscience and Biobehavioral Reviews* **3**, 301–341.

2 Sexual Differentiation and Neuronal Plasticity of Synaptic Organization During and After the Brain Development in the Rat

Masako NISHIZUKA

Department of Anatomy, Juntendo University School of Medicine, Tokyo 113

Mammalian sex determination appears to be genetically one of the developmental processes. Sexual differentiation in mammals begins at fertilization with the production of two types of embryos differing in chromosomal content. The Y chromosome obviously carries structural loci directing the formation of testes. Although sequencing events initiated by the expression of the gene are not definitively elucidated, recent studies have shown the localization of the testis-determining factor gene loci on a distal portion of the short arm of this chromosome (Sinclair *et al.*, 1990), whereas in the absence of the gene ovaries develop (George & Wilson, 1988).

Testosterone (T) produced by the testis has a major role in the growth and differentiation of many tissues in reproductive and non-reproductive organs as a result, whereas estrogens such as estradiol (E_2) secreted by ovaries have a major differentiating effect on the female reproductive tract, but they have minor effects on other organs (Gordon & Ruddle, 1981). T is the major hormone responsible for the sexual dimorphism of non-reproductive organs.

Gonadal steroids such as T, dihydrotestosterone (DHT), E, and progesterone (P) target the central nervous system (CNS) that takes an integral role in reproduction. In many mammalian species marked sexual dimorphisms exist in the brain and spinal cord. Involvements of sexually dimorphic and sexually differentiating parameters of the CNS have been documented in reproduction such as the regulation of gonadotropin (GTH) secretion and sexual behavior. The phenotypic sex of the CNS depends on gonadal sex in normal development.

Occurrence of sexual differentiation can be seen in each step of development of the CNS, such as in the neurogenesis (Jacobson & Gorski, 1981), cell migration (Breedlove & Arnold, 1981; Gorski, 1984; Sengelaub & Arnold, 1986, 1989), assembly of neurons (Gorski *et al.*, 1978; Breedlove & Arnold, 1980, 1981; Bleier *et al.*, 1982; Jordan *et al.*, 1982; Breedlove, 1984; Commins & Yahr, 1984; Swaab & Fliers, 1985; Ito *et al.*, 1986; Tobet *et al.*, 1986; Vigliettii-Panzica *et al.*, 1986; Allen *et al.*, 1989), selective neuronal death (Nordeen *et al.*, 1985; Murakami & Arai, 1989; Sengelaub & Arnold, 1989; Sengelaub *et al.*, 1989), growth of dendrites (Greenough *et al.*, 1977; Uchibori & Kawashima, 1985; Nishizuka *et al.*, 1989) and axons (Toran-Allerand, 1976; Loy & Milner, 1980), formation of neuronal networks (Raisman & Field, 1971, 1973; Dyer *et al.*, 1976; Matsumoto & Arai, 1980, 1986; Nishizuka & Arai, 1981a, 1982, 1983a, b; Dyer, 1984; Chen *et al.*,1990) and underlying cell to cell communication (Arimastu, 1981), expression of certain proteins (Stanley & Fink, 1986; Tobet & Fox, 1989), neurotransmitters (Simerly *et al.*, 1985a, b) and neuropeptides (de Vries *et al.*,1981; Wray & Hoffman, 1986; Micevych *et al.*, 1987a, b; Simerly & Swanson, 1987a, b), and in the neuronal plasticity (Loy & Milner, 1980; Matsumoto & Arai, 1981; Carrer & Aoki, 1982; Clough & Rodriguez-Sierra, 1983; Miyakawa & Arai, 1987; Nishizuka & Pfaff, 1989a; Gould *et al.*, 1990).

ROLE OF EARLY ANDROGEN SECRETION

The intrinsic pattern of brain sexuality in mammals is female, and a masculine pattern is brought about as a result of exposure to testicular hormones during early development; the inherent feminine development, masculinization, and defeminization are hence included in the sexual differentiation of the CNS. The typical female phenotype of brain sex differentiation, that is, the cyclic pattern of GTH secretion occurring in the mature female, is permanently suppressed and a tonic release pattern that characterizes the mature male brain is, in turn, elicited in neonataly T-treated female rats. A neonatal orchidectomy develops an inherently programmed feminine CNS. The sex differentiation of the CNS starts in the late prenatal period and finishes within a few days after birth in the rat, and phenotypes are, accordingly, expressed after puberty.

The processes of masculinization and those of defeminization seem to occur independently in the CNS. According to Yamanouchi *et al.*(1985) a fundamental architecture of the neuronal circuitry which drives masculine sexual behavior exists in the CNS of both the male and female rat and, similarly, the circuitry which drives the feminine behavior exists in the female and male CNS. In their experimental paradigm, the dorsal descending fiber connections which project to the preoptic area (POA) have been transected. Following castration and a replacement with E and P, these rats have shown a high lordosis quotient (LQ). The fact that the LQ of the male rats of which dorsal fibers have been transected is comparable to that of normal female indicates that a signal passing through the dorsal fibers are suppressing the activity of the lordosis driving neurocircuitry of the normally masculinized male. The female rats of which dorsal fibers is disconnected obtained a high LQ score when primed with a subthreshold dose of E, whereas normal female rats require a high dose of E to gain a similar level. Furthermore, the same transection facilitates the occurrence of masculine behavior in ovariectomized and T-treated rats (Yamanouchi, 1980). One can conclude, therefore, that the dorsal descending inputs of the POA of the female exert an inhibitory influence on the female sexual behavior and suppress the male behavior-mediating system existing in the female.

Neonatal administration of estrogens masculinizes certain functions of the CNS in the rat. Presence of aromatizing enzymes that transform testicular and exogenous T to E explains such a paradox (Naftolin & Ryan, 1975; Naftolin *et al.*, 1975; McEwen *et al.*, 1977). The aromatized E interacts with its receptor, thereby exerting its biological action. Therefore, the active form of the sex steroid in the sexual differentiation seems to be E. The enzymes are found in the CNS of the human (Naftolin *et al.*, 1975), rats (McEwen *et al.*, 1977), rabbits (George *et al.*, 1978) and quails (Schumacher & Balthazart, 1986). Such *in situ* formation theory is in good accordance with the findings that the non-aromatizable androgens have failed in brain masculinization (Arai, 1972; McDonald & Doughty, 1972).

If E is the active steroid in the sexual differentiation, the CNS of female fetuses must be protected from the maternal and placental estrogens. α-fetoprotein which is synthesized in the fetal liver and circulates in a perinatal period is known to bind to E but not to T in rats or mice (Nunez *et al.*, 1976; MacLusky & Naftolin, 1981; McEwen, 1981). The circulating E in male and female rat pups is captured by the protein, but the T can be transformed by the central aromatase into E in male pups. The immuno-reactivity against the fetoprotein is widely distributed in developing CNS, but the immunoreaction has been invisible in the hypothalamic and limbic neurons (Toran-Allerand, 1982). Supposedly, lack of the fetoprotein enables the aromatized E to reach its receptor in the cell nuclei of the hypothalamic and limbic neurons.

SEXUALLY DIMORPHIC CELL GROUPS IN THE CNS

Sexually dimorphic regions are distributed in forebrain, hindbrain and spinal cord of the rat, mouse (Bleier *et al.*, 1982), hamster (Greenough *et al.*, 1977), Mongolian gerbil (Commins & Yahr, 1984), ferret (Tobet *et al.*, 1986), cat (Calaresu & Henry, 1971), bird (Breedlove, 1984), monkey (Bubenlk & Brown, 1973; Ayoub *et al.*, 1983), and human (Swaab & Fliers, 1985, Allen *et al.*, 1989).

The main regions in rats are the septum (De Vries *et al.*, 1981), the bed nucleus of the stria terminalis (BST, Del Abril *et al.*, 1987; Simerly & Swanson, 1987a), the medial POA (Raisman & Field, 1971; Dyer *et al.*, 1976), the cell mass in the medial POA called the sexually dimorphic nucleus (SDN, Fig.1) in the POA (Gorski *et al.*, 1978), the anteroventral periventricular nucleus (AVPV, Fig. 2) of the POA (Bleier *et al.*, 1982; Simerly *et al.*, 1985a, b), the suprachiasmatic (Bleier *et al.*, 1982), ventromedial (VMH, Matsumoto & Arai, 1986) and arcuate nuclei (ARC, Matsumoto & Arai, 1980) of the hypothalamus, the hippocampus (Loy & Milner, 1980; Gould *et al.*, 1990), the medial amygdaloid nucleus (MAN, Fig. 3, Nishizuka & Arai, 1981a, 1983a, b), the cerebral cortex (Diamond, 1987), motor nuclei such as the hypoglossal nucleus (Yu, 1982, 1989) and the spinal nucleus of the bulbocavernosus (SNB, Breedlove & Arnold,

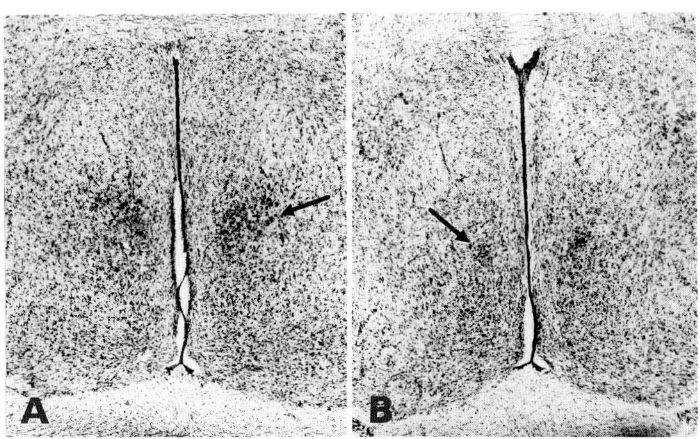

Fig. 1. Frontal section of the SDN-POA (arrows) of normal adult male (A) and female (B) rats in Nissl preparation. (Courtesy of S. Murakami).

1981) and the dorsolateral nucleus (DLN) in the lumbar spinal cord (Sengelaub & Arnold, 1989). The sex differentiation of these regions is primarily dependent on the early steroid action. Additionally, the sexual

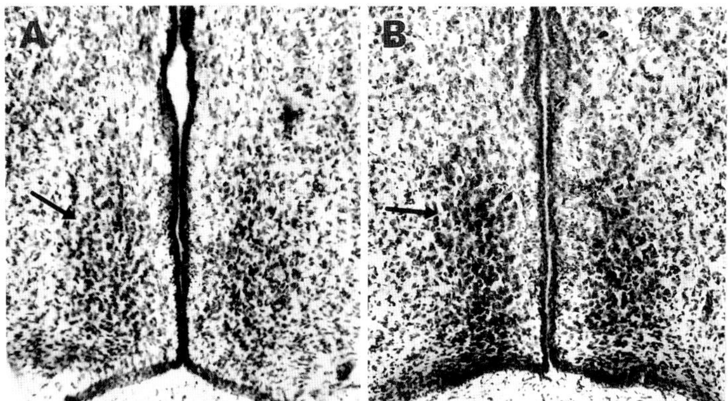

Fig. 2. Frontal section of the AVPV (arrows) of normal adult male (A) and female (B) rats in Nissl preparation. (Courtesy of S. Murakami).

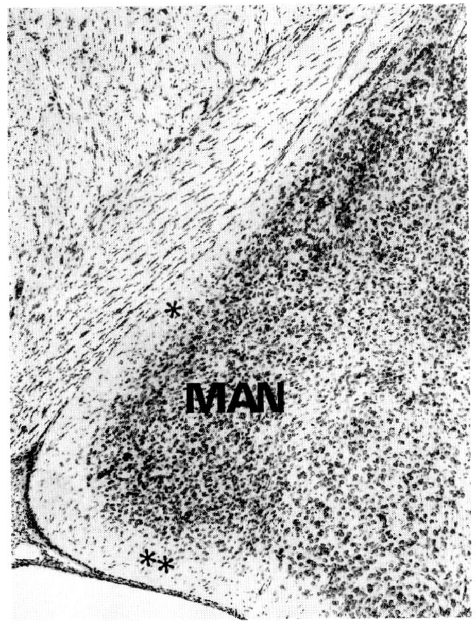

Fig. 3. Nissl preparation of the MAN. The region medial (*) and ventral (* *) to the cellular part (MAN) is the molecular layer.

dimorphisms are seen in pre- (Nadelhaft & McKenna, 1987) and post-ganglionic autonomic neurons (Wright, 1987).

Most of the sexually dimorphic brain and spinal regions notably play essential or regulatory roles in the GTH secretion (see Chapter 6) or the drive of sexual behavior (see Chapter 7). Moreover, these regions consist of the neurons that express androgen- and/or estrogen receptor molecules (Pfaff & Keiner, 1973; Stumpf & Grant, 1975; Simerly *et al.*, 1990; see Chapter 3) in addition to steroid-insensitive neuronal groups. Since the expression of these receptors begins in the late prenatal days in the rat CNS (Sheridan *et al.*, 1974; Weisz & Ward, 1980), the steroid-receptor complex undoubtedly plays a critical role in the brain sexual differentiation, whereas other neuroactive substances, *e.g.*, the nerve growth factor (Yanase *et al.*, 1988; Nishizuka & Arai, 1990), may play a secondary role.

MODIFICATION OF GENESIS, MIGRATION, AND DEATH OF THE CELL

In most of the brain regions of rats, the formation of neuronal cells is completed before birth. Biosynthesis of steroid receptors is thought not to be activated in undifferentiated neuroblasts but, rather, to occur after the final cell division in the neurogenesis. This probably implies that the participation of androgen or estrogen in the differentiation begins after the neurogenesis.

Figure 1 indicates that the SDN-POA of male rats consists of a greater number of neurons than that of females (Gorski *et al.*, 1978). Most neurons that eventually construct the SDN are formed in the germinal layer during 14–18 days of gestation (Jacobson & Gorski, 1981). When the date of birth was determined by the incorporation of ^{3}H-thymidine, the incidence of the SDN neurons that incorporated the isotope on day 17 in the male was greater than the female. It is still unclear whether the formation of SDN neurons on day 17 is reduced in female fetuses or if it continues, but some of the neurons that were formed on day 17 are eliminated during or after the migration in the female. The findings that a perinatal T treatment on or after day 18 of gestation has increased the size of the female SDN (Rhees *et al.*, 1990) but T given on or before day 17 has no effect imply that a population of late arising SDN neurons is present in both the male and the female. Functional significance of the SDN-POA has not been fully investigated (Arendash & Gorski, 1983; Anderson *et al.*, 1986).

The AVPV (Fig. 2) is another sexually dimorphic area in the POA. Since destruction of the AVPV region has suppressed the cyclic ovulation, the neuronal circuit of the AVPV is considered to play a role in the cyclic release of GTH in the female (Terasawa *et al.*, 1980). As shown in Fig. 2, the AVPV of female rats occupies a larger area and cells are distributed more densely than the male (see Bleier *et al.*, 1982). The nuclear size of the AVPV is, like the SDN, dependent on the level of gonadal steroid(s) in the perinatal period, but T given neonatally or late prenatally has decreased its size (Ito *et al.*, 1986). The AVPV is distinguished as a small cell cluster of periventricular neurons from the surrounding area in neonatal rats. A number of AVPV cells degenerate in male and female rats during postnatal development. Such cell death normally occurs during the first postnatal week and diminishes in the next week (Murakami & Arai, 1989). In female rats given 50 μg T-propionate(TP) on day 5, the incidence of pycnosis on day 7–10 has increased (Fig. 4, see Murakami & Arai, 1989). Furthermore, the number of picnosis on day 10 in normal males was greater than in normal females (Fig.4).

In the spinal motor nuclei called the SNB and DLN, that innervate striated penile muscles in the male, the early androgen modifies both the degeneration and migration of motoneurons. Motoneuron death occurring normally in the SBN is reduced in rats treated with T (Nordeen *et al.*, 1985; Sengelaub & Arnold, 1986, 199).

Supposing cell death were facilitated in the AVPV and suppressed in the spinal nuclei during sexual differentiation; the intracellular action of the steroids would then differ in individual cell groups. The possibility remains in the spinal nuclei that the androgen inhibits motoneuronal migration (Sengelaub & Arnold, 1986, 1989).

The AVPV neurons contain various neuropeptides (Kawata *et al.*, 1985; Simerly *et al.*, 1987b), dopamine or its synthesizing enzyme, tyrosine

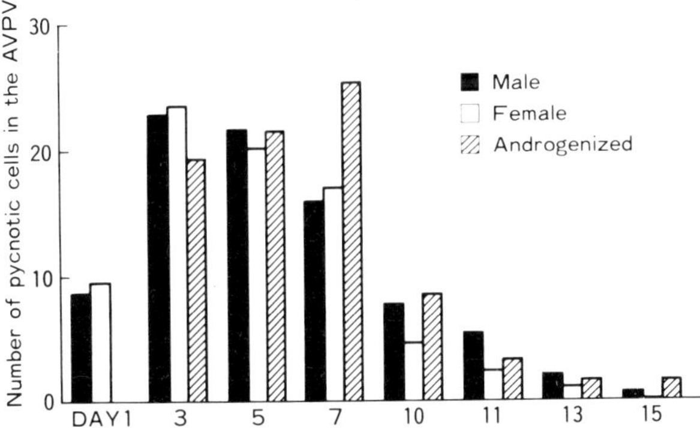

Fig. 4.   Number of pycnotic cells in the AVPV of normal male and female rats and androgenized rats receiving TP 50 μg on postnatal day 5. (Courtesy of S. Murakami).

hydroxylase (TH, Simerly *et al.*, 1985a). Dramatic sexual dimorphisms have been found in the distribution of leucine-enkephalin(ENK)- and TH-containing cells. The AVPV of the female contains 3–4 times more TH-containing cell bodies (Simerly *et al.*, 1985a) and fewer ENK-containing perikarya (Simerly *et al.*, 1988) than that of the male. The cell elimination (either the enhancement of cell death or a modulation of the cell migration) occurring as a process of sexual differentiation is obviously participating in the regulation of cell populations. The findings are noteworthy that the number of luteinizing hormone-releasing hormone (LHRH)-containing neurons is greater in female rats than in males (Wray & Hoffman, 1986).

SEXUAL DIMORPHISM IN SYNAPTIC ORGANIZATION

Figure 5 summarizes the neuronal connections, which are seen in a representative sexually dimorphic region (see below) called the MAN. Figures

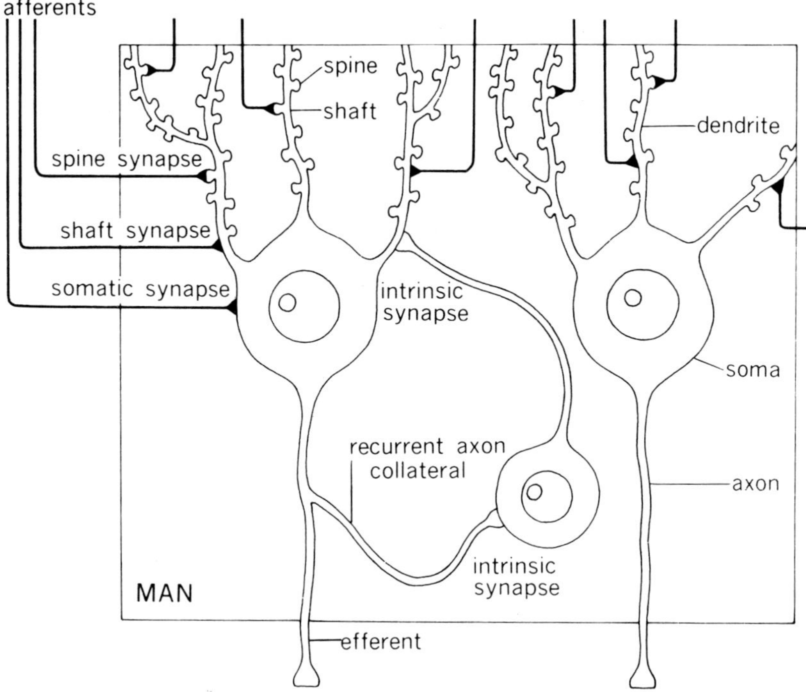

Fig. 5.   Neuronal connections in the MAN. Some representative neurons are shown. (Illustrated by S. Yasutomi).

6-9 show the ultrastructural feature of synapses. Dendrites of neurons provide the sites for synaptic transmission. The shafts (Fig. 6) of the dendritic trees and the spinous protrusions on the shafts (Fig. 8) have postsynaptic specialization and form junctions with presynaptic elements. They are the dendritic shaft synapses (Fig. 6) and the dendritic spine synapses (Fig. 8), respectively. Neuronal soma also forms synaptic connections (Fig. 7), however, the incidence of the somatic synapses seems low among synaptic populations overall. Most of the presynaptic elements are axons, but synapses between two dendrites (dendro-dendritic, Fig. 9) or two axons (axo-axonic) are plausible.

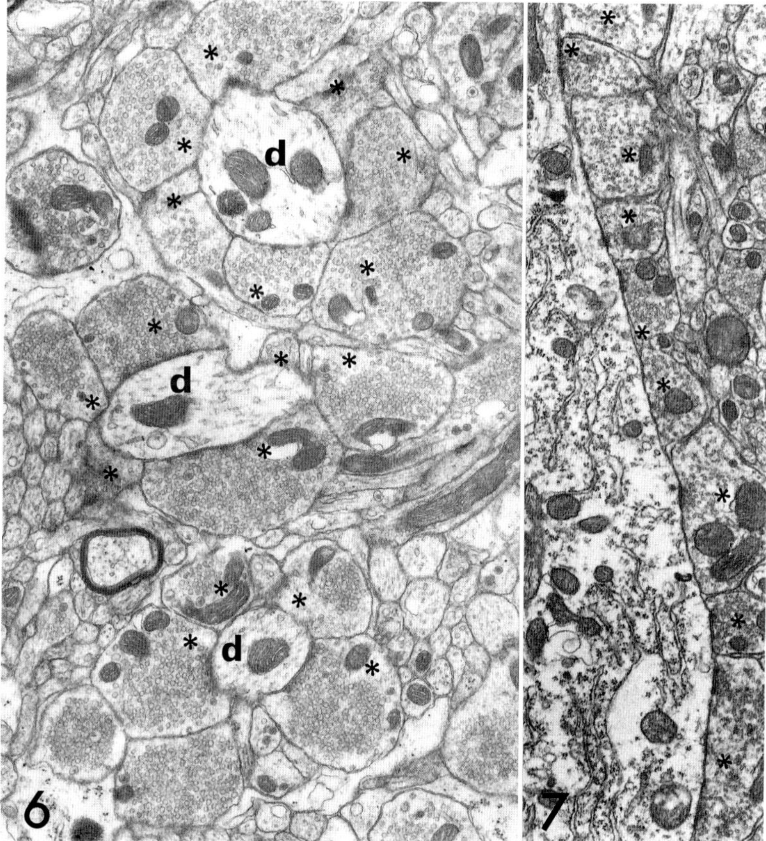

Figs. 6, 7.   Electron micrographs of the POA of adult female rats. (From Nishizuka & Pfaff, 1989b). Note the surface of dendritic shafts (d, Fig. 6) and a soma (Fig. 7) is covered by presynaptic terminals (*).

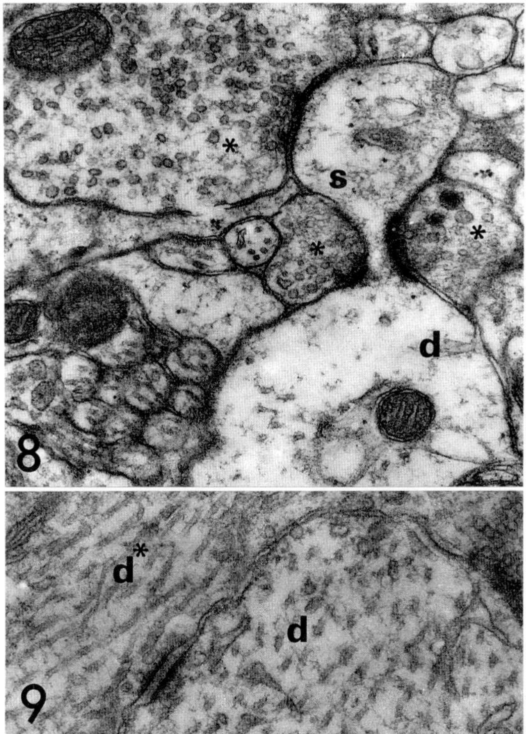

Figs. 8, 9.   Dendritic synapses in the POA of female rats. Fig. 8: dendritic spine (s) protrudes from dendritic shaft (d) and presynaptic terminals (∗) contact the spine and its stem. (From Nishizuka & Pfaff, 1989b). Fig. 9: presynaptic dendrite (d∗) synapsing on postsynaptic dendrite (d).

A pioneering observation on synaptic sexual dimorphisms was carried out by Raisman & Field (1971). A small area of the dorsal POA, the functional significance of which has not even yet determined (Brown-Grant *et al.*, 1977), receives the afferent arising from the amygdala and some other regions. Quantitative electron microscopic analyses revealed a sexual difference on the dendritic spines of the area. The density of the spine synapses formed with axons of non-amygdaloid origin(s) in the male was greater than that in the female (Raisman & Field, 1971, 1973). Neonatal orchidectomy and androgenization changed the sexuality that became apparent at adulthood (Raisman & Field, 1973).

Firing of LHRH neurons and the resultant LHRH release stimulates the release of GTH. The question whether the inputs converging into the LHRH neurons are sexually dimorphic has been answered. Recent electron microscopic analysis (Chen *et al.*, 1990) has revealed that the incidence of synapses on the LHRH neurons of female rats is greater than the male. Moreover, the female LHRH neurons have received β-endorphin-containing terminals more frequently than those of the male (Chen *et al.*, 1990). It is noteworthy that the LHRH neurons themselves are not the target for sex steroids because of lack of steroid receptors (Shivers *et al.*, 1983). To function as the component of the final common pathway, the LHRH neurons have to receive humoral information indirectly *via* the neural transmission. Most of the POA neurons receive a considerable number of synaptic inputs, as judging by the ultrastructure in which many terminals converge at the surface of neuronal soma (Fig. 7) and dendrites (Fig. 6, Nishizuka & Pfaff, 1989b). The density of synapses on the LHRH neurons is, in contrast, fairly low (Witkin & Silverman, 1985; Silverman 1988). This finding implies that the neural and humoral inputs have been mostly integrated in neurons presynaptic to the LHRH neurons, and that the sexually different LHRH release is hence attributable to the difference of connectivity in the LHRH neurons. The LHRH neurons participate in their own integration *via* the axons synapsing on them (Leranth *et al.*, 1985c).

SEXUAL DIFFERENTIATION IN THE SYNAPTOLOGY OF THE MEDIAL AMYGDALA

We have investigated the sexual differentiation of the synaptic organization in the amygdala of rats (Nishizuka & Arai, 1981a, b, 1982, 1983a–c). The

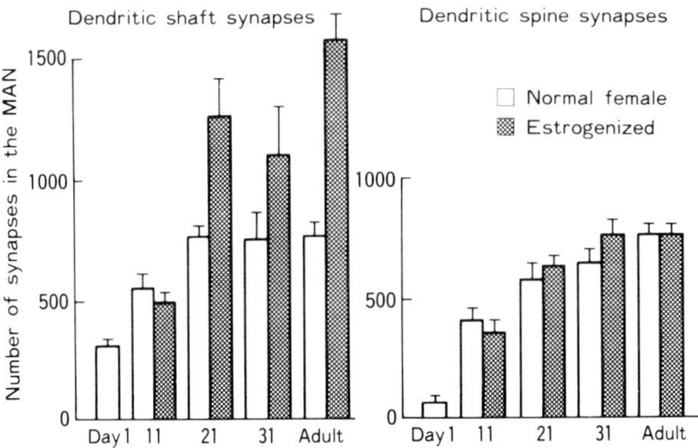

Fig. 10. Synaptogenesis in the MAN of normal and estrogenized female rats. E-benzoate was given for the first 30 days. The number of shaft and spine synapses (mean ± S.E.M. in a field of 10,000 sq. μm) are differentially counted. (From Nishizuka & Arai, 1981b).

MAN (Fig. 3) is known to be one of the steroid-target regions. As for its functional significance, earlier study showed that an electrical stimulation of the MAN promoted GTH secretion (Beltramino & Taleisnik, 1978), and that involvement of the MAN in regulation of male sexual behavior (Kondo & Arai, 1991) has been indicated. Figure 10 summarizes the progress of synapse formation in the MAN of female rats (Nishizuka & Arai, 1981b). The synaptogenesis begins before birth and finishes by the end of the third postnatal week. The number of synapses in the adult male is greater than in the adult female (Fig. 11, Nishizuka & Arai, 1981a), whereas the male and female MAN contains a comparable number of synapses during the second postnatal week (Fig. 12, Nishizuka & Arai, 1981a). The dimorphism in adult rats is due to the difference in the incidence of dendritic shaft synapses (Fig. 11). The shaft synapses in the

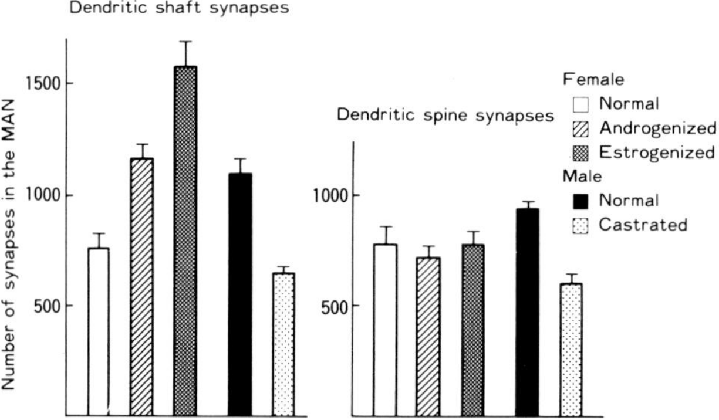

Fig. 11.   Number of synapses (mean ± S.E.M. in a field of 10,000 sq. μm) in the MAN of adult rats. Normal male and female rats, male rats castrated on the day of birth, and female rats given TP on day 5 (androgenized) and those given E-benzoate 30 days after birth (estrogenized) were studied. (From Nishizuka & Arai, 1981a, b).

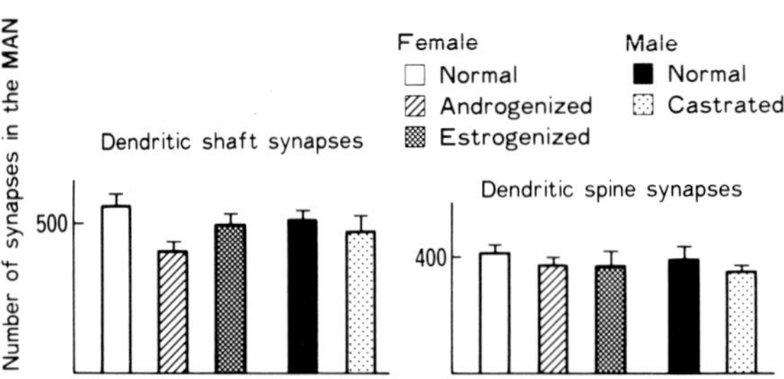

Fig. 12.   Number of synapses in the MAN of rats at 11 days of age. See the legend for Fig. 11.

male are more numerous than those of females, but the number of spine synapses in the two sexes is comparable (Fig. 11, Nishizuka & Arai, 1981a).

Androgen and estrogen both have potential to facilitate the formation of the dendritic synapses (Fig. 10, Nishizuka & Arai, 1981a, b). A continuous injection of E-benzoate for 30 postnatal days or a single injection of TP on day 5 has predominantly induced a shaft-synaptogenesis. The concomitant increase of the synapses occurs during the second and third weeks, which period corresponds to the final step of synaptogenesis in the MAN, even if the steroid treatment has been started earlier (Fig. 10). Thus there is a latency in the steroid-induced synaptogenic reaction.

Nonetheless, the sexual difference in the machinery for neural communication must be transmitted from presynaptic to postsynaptic neurons. An electrophysiological approach (Dyer *et al.*, 1976) detected a sex difference in the incidence of POA-anterior hypothalamic (AH) neurons that receive the amygdaloid inputs. In "endocrine male" rats, the POA-AH neurons projecting to the medial basal hypothalamus (MBH) receive the corticomedial amygdaloid fibers more frequently than observed in "endocrine females" (Dyer, 1984). The MBH consists of the VMH, ARC, and their adjacent regions.

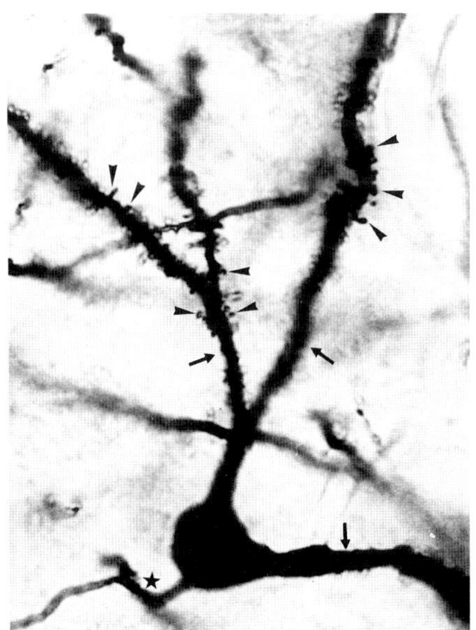

Fig. 13. A Golgi-impregnated neuron of the MAN of an adult female rat. Note dendrites (arrows) bearing dendritic spines (arrowheads). Axon (★) is visible.

In addition to the synapse density, the size of the MAN is sexually different, the male MAN being significantly larger (Mizukami *et al.*, 1983), and the total synapses in the MAN of adult male rats may hence be 1.4 times those of female MAN. One may speculate that the sexual difference in synapses simply reflects a difference in cell number. Our findings on sexual difference appearing on spines seem to provide evidence for an alternative possibility. We have recognized four groups of neurons in Golgi-impregnated preparations of the MAN (Nishizuka *et al.*, 1989). Among them, a subset of neurons (Fig. 13) is characterized by thick dendrites bearing numerous dendritic spines. The density of such spines in the male is significantly greater than in the female, indicating that the sexual differentiation of the MAN originates from the male-female difference in individual neurons.

The quantification of synapses has revealed that the incidence of spine synapses in the male is actually greater than in the female in the marginal zone of the MAN, so-called molecular layer (Fig. 3) (Nishizuka & Arai,

TABLE 1

Number of somatic synapses in the MAN of adult rats

	No. of rats	No. of neurons	No. of synapses/neuron (Mean ± S.E.M.)
Male			
Normal	5	210	2.2 ± 0.2
Castrated	4	164	2.6 ± 0.2
Female			
Normal	5	203	2.6 ± 0.2
Estrogenized	4	155	2.5 ± 0.1

Normal male and female rats, male rats castrated on the day of birth, and female rats given 50 μg TP on day 5 were studied. The number of synapses on the surface of each neuronal cell body on ultrathin section was counted under an electron microscope.

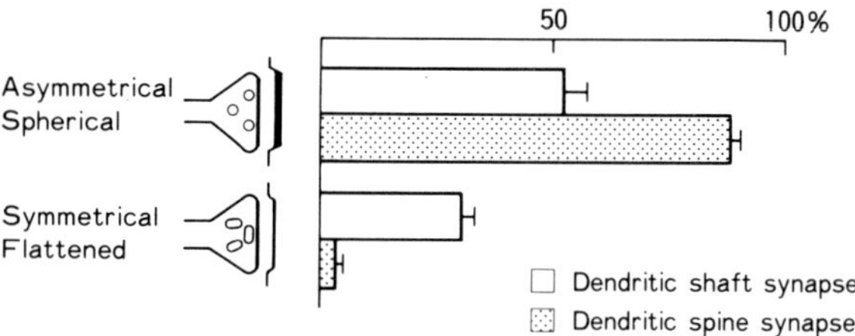

Fig. 14. Morphological features of synaptic junctions (asymmetrical or symmetrical) and synaptic vesicles (spherical or flattened) in the MAN. Mean frequency ± S.E.M. of three adult female rats is shown.

1983a). On the contrary, no sexual dimorphism has been found so far in somatic synapses of the MAN (Table 1).

Among synapses of the female MAN, more than half of the shaft synapses and 90 % of the spine synapses form the asymmetrical junction of which postsynaptic fibers have a marked membrane specialization and spherical synaptic vesicles (Fig. 14). They are excitatory in nature, depolarizing postsynaptic MAN neurons, whereas about 30% of the shaft synapses (Fig. 14) are inhibitory, being characterized by the symmetrical junctional complex and a few flattened vesicles in addition to the spherical ones.

GONADAL STEROIDS ACT ON PRE- AND POST-SYNAPTIC NEURONS

There is no doubt that both pre- and post-synaptic neurons cooperate in the enhancement of synapse formation in the masculinizing MAN. Such enhancement has been obtained similarly in MAN tissues that have developed in an ectopic site (Nishizuka & Arai, 1982). We have transplanted MAN tissues of neonatal rats into the eye chamber of adult rats, thereby eliminating afferently impinging presynaptic fibers. A continuous exposure to E has caused an increase of the density of shaft synapses in the transplants. This provides supportive evidence for the hypothesis that the postsynaptic neurons play a critical role in the reactive synaptogenesis. Arimatsu (1981) documented that the development of postsynaptic structure is directly regulated by the level of estrogen. The posterior part of the corticomedial amygdala of the mouse receives the cholinergic innervation and then expresses nicotinic acetylcholine receptor. The receptor activity measured by the binding capacity with α-bungarotoxin in the male has been greater than in the female. E has enhanced the appearance of the binding capacity in the posterior amygdalar neurons in a dissociated cell culture without any cholinergic innervation (Arimatsu *et al.*, 1985).

Even if the postsynaptic neurons would take a primary part, the participation of impinging axons should be taken into account. The synaptic organization of the molecular layer in the ventral and medial margin of the MAN has also been sexually dimorphic, its differentiation being dependent on the neonatal hormonal environment (Nishizuka & Arai, 1983a). The afferent fibers arising from the olfactory and accessory olfactory bulbs and intra-amygdaloid axonal connections from the posterior amygdala are main components of the presynapses of this layer. When we have made a lesion in the origin of the intra-amygdaloid input in male and female rats, a considerable number of synaptic terminals in the molecular layer have degenerated (Fig. 15), and then the synapses made by

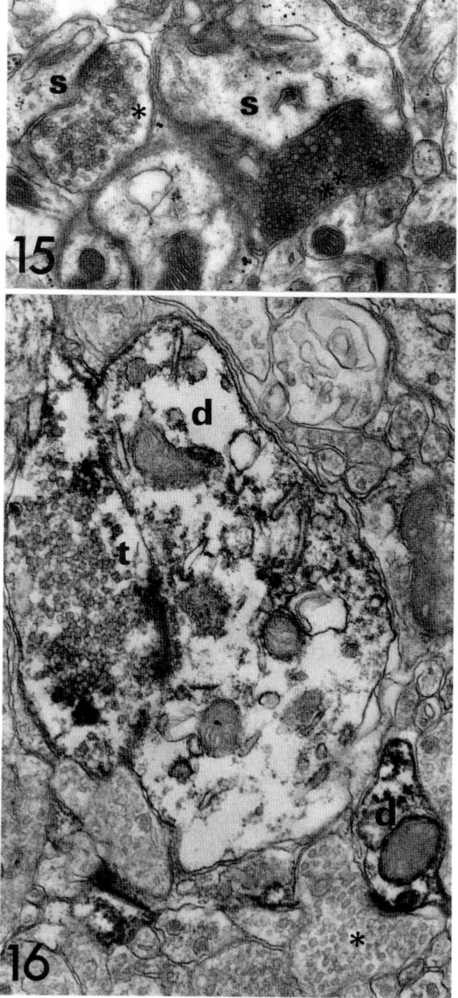

Fig. 15.   Synaptic profile of the molecular layer of the MAN following the lesion of the posterior amygdala. Note normal presynaptic terminal (*) synapsing on spine(s) and degenerating terminal (* *) on another spine. (From Nishizuka & Arai, 1983b).

Fig. 16.   Prolactin-like immunoreactive terminal (t) synapsing on prolactin-like immunoreactive dendrite (d) in the MBH of a rat. Immunonegative presynaptic terminal(*) synapses on another immunoreactive dendrite. (From Nishizuka et al., 1990).

these degenerating elements have disappeared. In the lesioned rats, the male-female differences of synaptology which are originally detected in the molecular layer have been invisible (Nishizuka & Arai, 1983b). These findings indicate that a substantial number of synaptic connections are made between the MAN neurons and the intra-amygdaloid axonal connection originating from or passing through the lesioned area. Further, the amount of the intra-amygdaloid connections differs between male and female rats. Occurrence of a concomitant promotion of axonal growth has been found in the mouse POA-hypothalamus cultured in E-containing media (Toran-Allerand, 1976). The neurons of the posterior corticomedial amygdala and those of the MAN possess the sex steroid receptors and, consequently, we cannot determine the site of steroid action (presynaptic, postsynaptic or both). The pre- and postsynaptic neurons seem to be activated simultaneously by the gonadal steroids.

INVOLVEMENT OF INTRINSIC SYNAPSES

I mentioned the synapse formation in the MAN transplanted in the eye in the previous section. The synapses formed in the transplants obviously connect two MAN neurons. A complete circumscribing deafferentation of the MAN has caused a marked degeneration of synaptic structures because all of the presynaptic axons converging to the MAN have been transected. In contrast, numerous synapses that have survived the circumscribing cut are considered to be the connections between two MAN neurons, and consequently intrinsic to the MAN. Thus, among synaptic populations in the MAN *in situ*, more than half seem to be "intrinsic synapses" (Nishizuka & Arai, 1983c). The presynaptic elements of these intrinsic synapses would be any short axon neurons or recurrent collaterals of the projection neurons (see Fig. 5).

The synaptic organization of the ARC and the VMH is also different between male and female rats (Matsumoto & Arai, 1980, 1986). These hypothalamic nuclei contain numerous intrinsic synapses. The synapses surviving the complete circumscribing cut account for about 50% of the total synaptic population (Matsumoto & Arai, 1981, Nishizuka & Pfaff, 1989a). Yagi and Sawaki (1975, 1978) provided electrophysiological evidence for the recurrent collaterals of the tubero-infundibular system. Some neurons of the MBH make a neural pathway mediating the recurrent inhibition and excitation. Among the neuronal groups involved in such recurrent controls (Yagi & Sawaki, 1978), the neurons containing γ-aminobutyric acid (GABA) are known to be sensitive to estrogen (Leranth *et al.*, 1985).

A considerable number of neurons in the VMH and the ARC contain an immunoreactive prolactin-like substance. They express the receptor for estrogen (Shivers *et al.*, 1989), and their prolactin contents are regulated by a fluctuating level of estrogen (Harlan *et al.*, 1989). The fiber which contains prolactin and projects into the midbrain apparently participates in the regulation of the lordosis behavior (Harlan *et al.*, 1983). We have found the prolactin-containing neurons synapsing on the dendrites of the prolactin-containing neurons (Fig.16, Nishizuka *et al.*, 1990). Since such connections are the synapses intrinsic to the sexually dimorphic regions, the possibility is raised that some of the intrinsic synapses form substantial amounts of sexually dimorphic microcircuits.

SEXUALLY DIMORPHIC NEURONAL NETWORKS

The neuronal subsets, the gross networks and the neuroactive substances which mediate neuronal activity seem to be common in the male and female neuronal networks. The functional sex difference of the CNS is not due to the presence of any gender-specific neuronal group or fiber tract. The process of sex differentiation in the synaptology may append an inhibitory or facilitatory circuit to an inherent framework.

Amounts of neuroanatomical studies have revealed the neuronal connections of the sexually dimorphic brain regions (see Palkovits & Zaborszky, 1979; Zaborszky, 1982; Swanson, 1987). These hodological findings enable us to conceive a chain of sexually dimorphic neurocircuits as follows: The MAN receives a sexually dimorphic input of intra-amygdaloid origin and sends the efferent fibers (see Fig. 5) to the sexually dimorphic cell groups such as the BST, POA and VMH. The sexually dimorphic nature of the connections between the MAN and the POA has been proved. The BST, POA, and the VMH reciprocally project to the MAN neurons. Particularly, the MBH neurons which project to the median eminence and a subset of POA neurons make a neural pathway converging to some of the amygdalar neurons (Hamamura & Yagi, 1980). The BST and the POA send reciprocal fiber connections. The POA mediates the recurrent control of the MBH neurons (Yagi & Sawaki, 1978). The sexually dimorphic forebrain regions are, consequently, interconnected. The sensory input which originates from the main and accessory olfactory bulbs impinges on the MAN and the BST. These regions mediate the neuronal transmission so that the olfactory information probably reaches to the POA, septum, and VMH.

In these forebrain key stations, the distribution of neurons containing cholecystokinin (CCK) is different between male and female rats

(Micevych *et al.*, 1987a, b; Simerly & Swanson, 1978b), and these CCK neurons may hence constitute a morphological and functional unit beyond neuroanatomical boundaries.

Among the POA neurons, the LHRH-neurons receive sexually different neural information (Chen *et al.*, 1990), and they transform it to sexual different humoral information *via* the median eminence and pituitary. In another sexually dimorphic cell cluster of the hypothalamus, the ARC, the GABA-neurons receive this humoral input and modify the activity of LHRH-neurons *via* a synaptic transmission (Leranth *et al.*, 1985a, b). Some of the LHRH-neurons may project to the midbrain central gray and stimulate the lordotic response (Sakuma & Pfaff, 1980; see Chapter 7). Some of the VMH-ARC neurons also project to the central gray and plays a key role in occurrence of lordotic response (Harlan *et al.*, 1983; Akaishi & Sakuma, 1986). The male-female difference in the MBH-midbrain projection has been proved (Sakuma, 1984). The septum and its efferent to the POA-hypothalamus also mediate the neuronal activity driving the male and female behaviors (Kondo *et al.*, 1990). The sexually dimorphic neuronal activity descends to the spinal cord and controls the skeletal muscles in mounting and lordosis.

In individual key stations of the male and the female, sexually different inputs, sexually common inputs and intrinsic synapses participate in modification, modulation, amplification, and synchronization of the neuronal activity. A feedback regulation *via* the reciprocal connections may join in such processing. Then the finally integrated neural activity is transmitted to the next stations. Estrogens and androgens play a critical part in all the station probably by changing the intracellular syntheses of certain molecules and/or acting on the neuronal membrane directly (Nabekura *et al.*, 1986).

SYNAPTIC PLASTICITY AND HORMONAL FEEDBACK

Manipulation of the steroid environments is required in neonatal period to reverse "endocrine sex" of the synaptology. This does not imply that the regulation of synaptic organization is exclusively exerted during the sexual differentiation. Recent studies have shown that a reorganization of the synaptic connections occurs in the brain of young and adult animals. In the rat ARC, the number of synapses increases abruptly at puberty. A rapid increase of estrogen which responds to the initiation of GTH secretion is responsible for such synaptogenesis (Matsumoto & Arai, 1977; Clough & Rodriguez-Sierra, 1983). The number of synapse in the ventrolateral part of the VMH decreased in ovariectomized adult rats as compared to rats

TABALE 2

Number (mean ± S.E.M./1000 sq. μm) of dendritic synapses in the VMH of adult female rats

		No. of rats	No. of shaft synapses	No. of spine synapses
VL	OVX	8	40.2 ± 1.7*	22.0 ± 2.0
	OVX-E	5	47.5 ± 2.0*	26.0 ± 2.6
DM	OVX	7	41.7 ± 1.9	20.7 ± 1.6
	OVX-E	5	45.9 ± 2.0	19.6 ± 2.6

*$P < 0.05$.

Rats were ovariectomized (OVX) 2 weeks before the analysis or ovariectomized and implanted with E-containing silastic capsule for 2 weeks (OVX-E). The ventrolateral (VL) and dorsomedial (DM) parts of the VMH were differentially studied. (From Nishizuka & Pfaff, 1989a).

ovariectomized but substituted by E, whereas the number in the steroid insensitive dorsomedial VMH did not change (Table 2, Nishizuka & Pfaff, 1989a). A single injection of E-benzoate maintains or increases the synaptic density in the VMH of female rats (Carrer & Aoki, 1982). An injection of E has induced a reactive synaptogenesis in the lateral septum of the adult female, but not in adult male rats (Miyakawa & Arai, 1987).

These findings indicate that a population of synapses in the mature CNS is formed or abandoned in response to the level of estrogen, and implies that, in female rats, the hardware of the neuronal networks fluctuates during the normal 4 or 5 day-estrous cycle. According to Gould *et al.* (1990), the dendritic spines of the hippocampal CA1 pyramidal neurons of adult female rats are reduced in number by ovariectomy. A replacement with E has restored the spine density within a few days. Moreover, an addition of P has intensely enhanced the formation of spines within a few hours. There is no doubt that the feedback of the gonadal steroids is responsible for an ongoing remodeling of connectivity in the neuronal networks. The feminine networks which have removed themselves from the activational action of the steroids in the neonatal days are conceivably under the rule of the gonads for life.

The synaptic remodeling occurring in the mature brain in response to fluctuation of the internal environments is a good example of the neuronal plasticity. Such dynamic property of neurons is the basis of the sexual differentiation and of the feedback action of gonadal steroids. In the synaptic reorganization, substrates which construct the presynaptic, post-synaptic and spinous membranes, cytoskeletal proteins and intercellular matrices might be supplied in individual cells at adulthood as well as in the neonatal days. Apart from these, a specific intracellular engram might be produced in immature neurons when they are or are not affected by the gonadal steroids. Little is known, however, about the primary intracellular event occurring in the sexual differentiation of the CNS.

CONCLUSION

The sexual differentiation of the CNS is triggered by the presence or absence of early steroid action. Existence of sexually dimorphic parameters has been documented in the various brain regions. The sexually dimorphic cell groups in the POA, hypothalamus, limbic area, brain stem, and the spinal cord form a chain of sexually differentiated and intensely interconnected neuronal networks.

It is conceivable that the neuronal subsets, fiber connections, and machinery which mediates the neural and humoral inputs are essentially common in the masculine and feminine neuronal networks.

Among the sexually dimorphic parameters, a male-female difference of the MAN appears in the synaptic organization. Postsynaptic MAN neurons seem to play a critical part in the synaptic masculinization. On the other hand, involvement of the intra-amygdaloid axonal connections converging on the MAN is also essential. In the sexually different synaptic connections, both presynaptic and postsynaptic neurons are presumably sensitive to the steroids, probably cooperating during sexual differentiation. The key stations of the sexually dimorphic networks take part in reproduction such as the regulation of the GTH secretion and the driving of male and female behavior. Large amounts of intrinsic synapses that exist in the sexually dimorphic regions are involved in modification and integration of the neural and humoral inputs.

Remodeling of neuronal connections occurs without a letup in response to fluctuating levels of the gonadal steroids. Such plastic property of neurons is the basis for the synaptic sexual differentiation and the hormonal feedback in the CNS.

REFERENCES

Akaishi, T. & Sakuma, Y. (1986). Projection of oestrogen-sensitive nucleus of the female rats. *Journal of Physiology (London)* **372**, 207-220.

Allen, L.S., Hines, M., Shryne, J. & Gorski, R.A. (1989). Two sexually dimorphic cell groups in the human brain. *Journal of Neuroscience* **9**, 497-506.

Anderson, R.H., Fleming, D.E., Rheeds, R.W. & Kinghorn, E. (1980). Relationships between sexual activity, plasma testosterone, and the volume of the sexually dimorphic nucleus of the preoptic area in prenatally stressed and non-stressed rats. *Brain Research* **370**, 1-10.

Arai, Y. (1972). Effect of α-dihydrotestosterone on differentiation of masculine pattern of the brain in the rat. *Endocrinologia Japonica* **19**, 389–393.

Arendash, G.W. & Gorski, R.A. (1983). Effects of discrete lesions of the sexually dimorphic nucleus of the preoptic area or other medial preoptic regions on the sexual behavior of male rats. *Brain Research Bulletin* **10**, 147–154.

Arimatsu, Y. (1981). Sexual dimorphism in α-bungarotoxin binding capacity in the mouse amygdala. *Brain Research* **213**, 432–437.

Arimatsu, Y., Kondo, S. & Kojima, M. (1985). Enhancement by estrogen treatment of α-bungarotoxin binding in fetal mouse amygdala cells aggregates *in vitro*. *Neuroscience Research* **2**, 211–220.

Ayoub, D.M., Greenough, W.T. & Juraska, J.M. (1983). Sex differences in dendritic structure in the preoptic area of the juvenile macaque monkey brain. *Science* **219**, 197–198.

Beltramino, C. & Taleisnik, S. (1978). Facilitatory and inhibitory effects of electro-chemical stimulation of the amygdala on the release of luteinizing hormone. *Brain Research* **144**, 95–107.

Bleier, R., Byne. W. & Siggelkow, I. (1982). Cytoarchitectonic sexual dimorphisms of the medial preoptic and anterior hypothalamic areas in guinea pig, rat, hamster and mouse. *Journal of Comparative Neurology* **212**, 118–130.

Breedlove, S.M. (1984). Androgen forms sexually dimorphic spinal nucleus by saving motoneurons from programmed death. *Society for Neuroscience Abstracts* **10**, 927.

Breedlove, S.M. & Arnold, A.P. (1980). Hormone accumulation in a sexually dimorphic motor nucleus of the rat spinal cord. *Science* **210**, 564–566.

Breedlove, S.M. & Arnold, A.P. (1981). Sexually dimorphic motor nucleus in rat spinal cord: Response to adult hormone manipulation, absence in androgen-insensitive rats. *Brain Research* **225**, 297–307.

Brown-Grant, K., Murray, M.A.F. Raisman, G. & Sood, M.C. (1977). Reproductive function in male and female rats following extra-and intra-hypothalamic lesions. *Proceedings of Royal Society, London,* **B198**, 267–278.

Bubenlk, G.A. & Brown, G.M. (1973). Morphologic sex differences in primate brain areas involved in regulation of reproductive activity. *Experientia,* **29**, 619–621.

Calaresu, F.R. & Henry, J.L. (1971). Sex difference in the number of sympathetic neurons in the spinal cord of the cat. *Science* **173**, 343–344.

Carrer, H.F. & Aoki, A. (1982). Ultrastructural changes in the hypothalamic ventromedial nucleus of ovariectomized rats after estrogen treatment. *Brain Research* **240**, 221–233.

Chen, W.P., Witkin, J.W. & Silverman, A.-J. (1990). Sexual dimorphism in the synaptic input to gonadotropin releasing hormone neurons. *Endocrinology* **126**, 695–702.

Clough, R.W. & Rodriguez-Sierra, J.F. (1983). Synaptic changes in the hypothalamus of the prepuberal female rat administered estrogen. *American Journal of Anatomy* **167**, 205–214.

Commins, D. & Yahr, P. (1984). Adult testosterone levels influence the morphology of sexually dimorphic area in Mongolian gerbil brain. *Journal of Comparative Neurology* **224**, 132–140.

Del Abril,A., Segovia, S. & Guillamon,A. (1987). The bed nucleus of the stria terminalis in the rat: regional sex differences controlled by gonadal steroids early after birth. *Developmental Brain Research* **32**, 295–300.

De Vries, Buijs, R.M. & Swaab, D.F. (1981). Ontogeny of the vasopressinergic neurons of the suprachiasmatic nucleus and their extrahypothalamic projection in the rat brain-presence of a sex difference in the lateral septum. *Brain Research* **218**, 67–78.

Diamond, M.C. (1987). Sex differences in the rat forebrain. *Brain Research Reviews* **12**, 235–240.

Dyer, R.G. (1984). Sexual differentiation of the forebrain-relationship to gonadotrophin secretion. *Progress in Brain Research* **61**, 223–236.

Dyer, R.G., MacLeod, N.K. & Ellendorf, F. (1976). Electrophysiological evidence for sexual dimorphism and synaptic convergence in the preoptic and anterior hypothalamus areas of the rat. *Proceedings of Royal Society, London* **B193**, 421–440.

George, F.W., Tobelman, W.T., Milewich, L. & Wilson, J.D. (1978). Aromatase activity in the developing rabbit brain. *Endocrinology* **102**, 86–91.

George, F.W. & Wilson, J.D. (1988). Sex determination and differentiation. In: *The Physiology and Reproduction*, Vol. 1 (Knobil, E. and Neil. J., eds.), Raven Press, New York, pp. 3–26.

Gordon, J.W. & Ruddle, F.H. (1981). Mammalian gonadal determination and gametogenesis. *Science* **211**, 1265–1271.

Gorski, R.A. (1984). Critical role for the medial preoptic area in the sexual differentiation of the brain. *Progress in Brain Research* **61**, 129–146.

Gorski, R.A., Gordon, J.H., Shryne, J.E. & Southam, A.M. (1978). Evidence for a morphological sex difference within the medial preoptic area of the rat brain. *Brain Research* **148**, 333–346.

Gould, E., Woolley, C.S., Frankfurt, M. & McEwen, B.S. (1990). Gonadal steroids regulate dendritic spine density in hippocampal pyramidal cells in adulthood. *Journal of Neuroscience* **10**, 1286–1291.

Greenough,W.T., Carter,C.S., Steerman, C. & DeVoogd,T.J. (1977). Sex differences in dendritic patterns in hamster preoptic area. *Brain Research* **126**, 63–72.

Hamamura, M. & Yagi, K. (1980). Amygdala neurons: converging synaptic inputs produced by median eminence and medial preoptic area stimulations in rats. *Journal of Physiology* **300**, 515–524.

Harlan, R.E., Shiver, B.D. & Pfaff, D.W. (1983). Midbrain microinfusions of prolactin increase the estrogen-dependent behavior, lordosis. *Science* **219**, 1451–1453.

Harlan, R.E., Shiver, B.D., Fox, S.R., Paplove, K.A., Schachter, B.S. & Pfaff, D.W. (1989). Distribution and partial characterization of immunoreactive prolactin in the rat brain. *Neuroendocrinology* **49**, 7–22.

Ito, S., Murakami, S., Yamanouchi, K. & Arai, Y. (1986). Perinatal androgen exposure decreases the size of the sexually dimorphic medial preoptic nucleus in the rat. *Proceedings of the Japan Academy* **62** Ser.B, 408–411

Jacobson, C.D. & Gorski, R.A. (1981). Neurogenesis of the sexually dimorphic nucleus of the preoptic area in the rat. *Journal of Comparative Neurology* **196**, 519–529.

Jordan, C.L., Breedlove, S.M. & Arnold, A.P. (1982). Sexual dimorphism and the influence of neonatal androgen in the dorsal motor nucleus of the rat lumbar spinal cord. *Brain Research* **249**, 309–314.

Kawata, M., Nakao, K., Morii, N., Kiso, Y., Yamashita, H., Imura, H. & Sano, Y. (1985). Atrial natriuretic polypeptide: Topographical distribution of the rat brain by radioimmunoassay and immunohistochemistry. *Neuroscience* **16**, 521–546.

Kondo, Y. & Arai, Y. (1991). Limbic regulation of copulatory behavior in male rats: a dual lesion study of the medial preoptic area and medial amygdala. *Neuroscience Research*, Suppl. 14, S103.

Kondo, Y., Shinoda, A., Yamanouchi, K. & Arai, Y. (1990). Role of septum and preoptic area in regulating masculine and feminine sexual behavior in male rats.

Hormone and Behavior **24**, 241–434.

Leranth, C., MacLusky, N.J., Sakamoto, H., Shanabrough, M. & Naftolin, F. (1985a). Glutamic acid decarboxylase containing axons synapse on LHRH neurons in the rat medial preoptic area. *Endocrinology* **40**, 536–539.

Leranth, C., Sakamoto, H., MacLusky, N.J., Shanabrough, M. & Naftolin, F. (1985b) Estrogen responsive cells in the arcuate nucleus of the rat contain glutamic acid decarboxylase (GAD): an electron microscopic immunocytochemical study. *Brain Research* **331**, 376–381.

Leranth, C., Segraum, L.M., Palkovits, M., MacLusky, N.J., Shanabrough, M. & Naftolin, F. (1985c). The LH-RH containing neuronal network in the preoptic area of the rat: Demonstration of LH-RH containing nerve terminals in synaptic contact with LH-RH neurons. *Brain Research* **345**, 332–336.

Loy, R. & Milner, T.A. (1980). Sexual dimorphisms in extent of axonal sprouting in rat hippocampus. *Science* **208**, 1282–1284.

MacLusky, N.J. & Naftolin, F. (1981). Sexual differentiation of the central nervous system. *Science* **211**,1294–1303.

Matsumoto, A. & Arai, Y. (1977). Precocious puberty and synaptogenesis in the hypothalamic arcuate nucleus in pregnant mare serum gonadotropin (PMSG) treated immature female rats. *Brain Research* **129**, 475–478.

Matsumoto, A. & Arai, Y. (1980). Sexual dimorphism in "wiring pattern" in the hypothalamic arcuate nucleus and its modification by neonatal hormone environment. *Brain Research* **190**, 238–242.

Matsumoto, A. & Arai, Y. (1981). Neural plasticity in the deafferented hypothalamic arcuate nucleus of adult female rats and its enhancement by treatment with estrogen. *Journal of Comparative Neurology* **197**, 197–205.

Matsumoto, A. & Arai, Y. (1986). Male-female difference in synaptic organization of the ventromedial nucleus of the hypothalamus in the rat. *Neuroendocrinology* **42**, 232–236.

McDonald, P.G. & Doughty, C. (1972). Comparison of the effects of neonatal administration of testosterone and dihydrotestosterone in the female rat. *Journal of Reproduction and Fertility* **30**, 55–62.

McEwen, B.S. (1981). Neural gonadal steroid action. *Science* **211**, 1303–1311.

McEwen, B.S., Lieberburg, I., Chaptal, C. & Krey, L.C. (1977). Aromatization: Important for sexual differentiation of the neonatal rat brain. *Hormone and Behavior* **9**, 249–263.

Micevych, P., Akesson,T. & Elde, R. (1987a). The distribution of cholecystokinin-immunoreactive cell bodies in male and female rat. I. Hypothalamus. *Journal of Comparative Neurology* **255**,124–136.

Micevych, P., Akesson, T. & Elde, R. (1987b). The distribution of cholecystokinin-immunoreactive cell bodies in male and female rat. II. Bed nucleus of the stria terminalis and amygdala. *Journal of Comparative Neurology* **269**, 381–391.

Miyakawa, M. & Arai, Y. (1987). Synaptic plasticity to estrogen in the lateral septum of the adult male and female rats. *Brain Research* **436**, 184–188.

Mizukami, S., Nishizuka, M. & Arai, Y. (1983). Sexual difference in nuclear volume and its ontogeny in the rat amygdala. *Experimental Neurology* **79**, 565–575.

Murakami, S. & Arai, Y. (1989). Neuronal death in the developing sexually dimorphic periventricular nucleus of the preoptic area in the female rat: effect of neonatal androgen treatment. *Neuroscience Letters* **102**, 185–190.

Nabekura, J., Oomura, Y., Minami, T., Mizuno, Y. & Fukuda, A. (1986). Mechanism of the rapid effect on 17,β-estradiol on medial amygdala neurons. *Science* **233**, 226-228.

Nadelhaft, I. & McKenna, K.E. (1987). Sexual dimorphism in sympathetic preganglion neurons of the rat hypogastric nerve. *Journal of Comparative Neurology* **256**, 308-315.

Naftolin, F. & Ryan, K.J. (1975). The metabolism of androgens in central neuroendocrine tissues. *Journal of Steroid Biochemistry* **6**, 993-997.

Naftolin, F. Ryan, K.J., Davis, I.J., Reddy, V.V. Flores, F., Petro, Z., Kuhn, M. White, R.J., Takaoka, Y. & Wolin, L. (1975). The formation of estrogen by central neuroendocrine tissues. *Recent Progress in Hormone Research* **31**, 295-315.

Nishizuka, M. & Arai, Y. (1981a). Sexual dimorphism in synaptic organization in the amygdala and its dependence on neonatal hormone environment. *Brain Research* **212**, 31-38.

Nishizuka, M. & Arai, Y. (1981b). Organizational action of estrogen on synaptic pattern in the amygdala: implications for sexual differentiation of the brain. *Brain Research* **213**, 422-426.

Nishizuka, M. & Arai, Y. (1982). Synapse formation in response to estrogen in the medial amygdala developing in the eye. *Proceedings of the National Academy of Sciences, U.S.A.* **79**, 7024-7026.

Nishizuka, M. & Arai Y. (1983a). Regional difference in sexually dimorphic synaptic organization of the medial amygdala. *Experimental Brain Research* **49**, 462-465.

Nishizuka, M. & Arai, Y. (1983b). Male-female differences in the intra-amygdaloid input to the medial amygdala. *Experimental Brain Research* **52**, 328-332.

Nishizuka, M. & Arai, Y. (1983c). Intrinsic connections in the medial amygdala as revealed by complete deafferentation. *Neuroscience Letters* **35**, 247-251.

Nishizuka, M. & Pfaff, D.W. (1989a). Intrinsic synapses in the ventromedial nucleus of the hypothalamus: An ultrastructural study. *Journal of Comparative Neurology* **286**, 260-268.

Nishizuka, M. & Pfaff, D.W. (1989b). Medial preoptic islands in the rat brain: electron microscopic evidence for intrinsic synapses. *Experimental Brain Research* **77**, 295-301.

Nishizuka, M. Murakami, S. & Arai, Y. (1989). Neuronal structure and male-female difference in the medial amygdala of rats. *Biomedical Research* **10**, S3, 323-327.

Nishizuka, M. & Arai, Y. (1990). Modification of nerve growth factor and sex steroids in development of hypothalamic neurons. *Endocrinologia Experimentalis* **24**, 77-86.

Nishizuka, M., Shivers, B.D., Leranth, C. & Pfaff, D.W. (1990). Ultrastructural characterization of prolactin-like immunoreactivity in rat medial basal hypothalamus. *Neuroendocrinology* **51**, 249-254.

Nordeen, E.J., Nordeen, K.W., Sengelaub, D.R. & Arnold, A.P. (1985). Androgens prevent normally occurring cell death in a sexually dimorphic spinal nucleus. *Science* **229**, 6712-6717.

Nunez, E.A., Benassayag, C., Savu, L., Valleette, G. & Jayle, M.F. (1976). Serum binding of some steroid hormones during development in different animal species. Discussion of the biological significance of this binding. *Annals of Biology, Animal Biochemistry and Biophysics* **16**, 491-501.

Palkovits,M. & Zaborszky,L. (1979). Neural connections of the hypothalamus. In: *Handbook of the Hypothalamus*, Vol. 1, Anatomy of the Hypothalamus (Morgane, P.J. and Panksepp, J., eds.), Marcel Dekker,Inc., New York, pp. 379-509.

Pfaff, D.W. & Keiner, M. (1973). Atlas of estradiol-concentrating cells in the central nervous system of the female rat. *Journal of Comparative Neurology* **151**, 121-158.

Raisman, G. & Field, P.M. (1971). Sexual dimorphism in the preoptic area of rats. *Science* **173**, 731-733.

Raisman, G. & Field, P. M. (1973). Sexual dimorphism in the neuropil of the preoptic area of the rat and its dependence on neonatal androgen. *Brain Research* **54**, 1-29.

Rhees, R.W., Shryne, J.E. & Gorski, R.A. (1990). Onset of the hormone-sensitive perinatal period for sexual differentiation of the sexually dimorphic nucleus of the preoptic area in female rats. *Journal of Neurobiology* **21**, 781-786.

Sakuma, Y. (1984). Influences of neonatal gonadectomy or androgen exposure on the sexual differentiation of the rat ventromedial hypothalamus. *Journal of Physiology* **349**, 273-286.

Sakuma, Y. & Pfaff, D.W. (1980). Effects of LHRH and antibody to LHRH infused in central grey on lordosis behavior in female rats. *Nature* **283**, 566-567.

Schumacher, M. & Balthazart, J. (1986). Testosterone-induced brain aromatase is sexually dimorphic. *Brain Research* **370**, 285-293.

Sengelaub, D.R. & Arnold, A.P. (1986). Development and loss of early projections in a sexually dimorphic rat spinal nucleus. *Journal of Neuroscience* **6**, 1613-1620.

Sengelaub, D.R. & Arnold, A.P. (1989). Hormonal control of neuron number in sexually dimorphic spinal nuclei of the rat: I.Testosterone-regulated death in the dorsolateral nucleus. *Brain Research* **280**, 622-629.

Sengelaub, D.R., Nordeen. E.J., Nordeen, K.W. & Arnold, A.P. (1989). Hormonal control of neuron number in sexually dimorphic spinal nuclei of the rat: III. Differential effects of the androgen dihydrotestosterone. *Journal of Comparative Neurology* **280**, 637-644.

Sheridan, P.J., Sar. M. & Stumps, W.E. (1974). Autoradiographic localization of ^{3}H-estradiol or its metabolites in the central nervous system of the developing rat. *Endocrinology* **94**, 1386-1390.

Shivers, B.D., Harlan R.E., Morrell, J.I. & Pfaff, D.W. (1983). Absence of estradiol concentration in cell nuclei of LHRH-immunoreactive neurons. *Nature* **304**, 345-347.

Shivers, B.D., Harlan, R.E. & Pfaff, D.W. (1989). A subset of neurons containing immunoreactive prolactin is a target for estrogen regulation of gene expression in rat hypothalamus. *Neuroendocrinology* **49**, 23-27.

Silverman, A.-J. (1988). The gonadotropin-releasing hormone (GnRH) neuronal systems: Immunocytochemistry. In: *The Physiology and Reproduction*, Vol. 1 (Knobil, E. and Neil, J., eds.), Raven Press, New York, pp. 1283-1304.

Simerly, R.B., Swanson, L.W., Handa, R.J. & Gorski, R.A. (1985a). Influence of perinatal androgen on the sexually dimorphic distribution of tyrosine hydroxylase-immunoreactive cells and fibers in the anteroventral periventricular nucleus of the rat. *Neuroendocrinology* **40**, 501-510.

Simerly, R.B., Swanson L. W.& Gorski, R.A. (1985b). The distribution of monoaminergic cells and fibers in a periventricular preoptic nucleus involved in the control of gonadotropin release: Immunohistochemical evidence for a dopaminergic sexual dimorphism. *Brain Research* **330**, 55-64.

Simerly, R.B. & Swanson, L.W. (1987a). Castration reversibly alters levels of cholecystokinin immunoreactivity within cells of three interconnected sexually dimorphic forebrain nuclei in the rat. *Proceedings of the National Academy of Sciences, U.S.A.* **84**, 2087-2091.

Simerly, R.B. & Swanson, L.M. (1987b). The distribution of neurotransmitter-specific cells and fibers in the anteroventral periventricular nucleus: Implications for the control of gonadotropin secretion in the rat. *Brain Research* **400**, 11–34.

Simerly. R.B., McCall, L.D. & Swanson, L.D. (1988). Distribution of opioid peptides in the preoptic region: immunocytochemical evidence for a steroid-sensitive enkephalin sexual dimorphism. *Journal of Comparative Neurology* **276**, 442–459.

Simerly, R.B., Chang, C. Muramastu, M. & Swanson, L.W. (1990). Distribution of androgen and estrogen receptor mRNA-containing cells in the rat brain: an *in situ* hybridization study. *Journal of Comparative Neurology* **294**, 76–95.

Sinclair, A.H., Berta, P., Palmer, M.S., Hawkins, J.R., Griffiths, B.L., Smith, M.J., Foster, J.W., Frischauf, A.M., Lovell-Badge, R. & Goodfellow, P.N. (1990). A gene from the human sex-determining region encodes a protein with homology to a conserved DNA-binding motif. *Nature* **346**, 240–244.

Stanley, H.F. & Fink, G. (1986). Synthesis of specific brain proteins is influenced by testosterone at mRNA level in the neonatal rat. *Brain Research* **370**, 223–231.

Stumpf, W.E. & Grant, L.D. (1975). *Anatomical Neuroendocrinology*, S. Karger, Basel.

Swaab, D.F. & Fliers, E. (1985). A sexually dimorphic nucleus in the human brain. *Science* **228**, 1112–1115.

Swanson,L. M. (1987). The hypothalamus. In: *Handbook of Chemical Neuroanatomy*, Vol. 5, Integrated Systems of the CNS, Part 1, Hypothalamus, Hippocampus, Amygdala, Retina (Björklund, A., Hökfelt, A. and Swanson, L.M., eds.), Elsevier, Amsterdam, pp. 1–124.

Terasawa, E. Wiegand, S.J. & Bridson, W.E. (1980). A role of the medial preoptic nucleus on afternoon of proestrus in female rats. *American Journal of Physiology* **238**, E533–539.

Tobet, S.A., Zahnister, D.H. & Baum, M.J. (1986). Sexual dimorphism in the preoptic/hypothalamic area of ferrets: effects of adult exposure to sex steroids. *Brain Research* **364**, 249–257.

Tobet, S.A. & Fox, T.O. (1989). Sex and hormone dependent antigen immunoreactivity in developing rat hypothalamus. *Proceedings of the National Academy of Sciences, U.S.A.* **86**, 382–386.

Toran-Allerand, C.D. (1976). Sex steroids and the development of the newborn mouse hypothalamus and preoptic area *in vitro*:implications for sexual differentiation. *Brain Research* **106**, 407–412.

Toran-Allerand, C.D. (1982). Regional differences in interneuronal localization of alpha-fetoprotein in developing mouse brain. *Developmental Brain Research* **5**, 213–217.

Uchibori, M. & Kawashima, S. (1985). Effects of sex steroids on the growth of neuronal processes in neonatal rat hypothalamus-preoptic area and cerebral cortex in primary culture. *International Journal of Developmental Neuroscience* **3**, 167–176.

Viglietti-Panzica, C., Panzica, G.C., Fiori, M.G., Calcagni, M., Anselmetti, G.C. & Abalthazart, J. (1986). A sexually dimorphic nucleus of the quail preoptic area. *Neuroscience Letters* **64**, 129–134.

Weisz, J. & Ward, I.L. (1980). Plasma testosterone and progesterone titers of pregnant rats, their male and female offspring. *Endocrinology* **106**, 306–316

Witkin, J.W. & Silverman, A.J. (1985). Synaptology of LHRH neurons in rat preoptic area. *Peptides* **6**, 263–271.

Wray, S. & Hoffman, G. (1986). A developmental study of the quantitative distribution of LHRH neurons within the central nervous system of postnatal male and female rats. *Journal of Comparative Neurology* **252**, 522–531.

Wright, L.L. (1987). Development of the sex difference in neuron numbers of the superior cervical ganglion: Effects of transection of the cervical sympathetic trunk. *Journal of Comparative Neurology* **263**, 259-264.

Yagi, K. & Sawaki, Y. (1975). Recurrent inhibition and facilitation: demonstration in the tubero-infundibular system and effects of strychnine and picrotoxin. *Brain Research* **84**, 155-159.

Yagi, K. & Sawaki, Y. (1978). Electrophysiological characteristics of identified tubero-infundibular neurons. *Neuroendocrinology* **26**, 50-64.

Yamanouchi, K. (1980). Mounting and lordosis behavior in androgen primed ovariectomized rats: Effect of dorsal deafferentation of the preoptic area and hypothalamus. *Endocrinologia Japonica* **27**, 499-504.

Yamanouchi, K., Matsumoto, A. & Arai, Y. (1985). Neural and hormonal control of lordosis behavior in the rat. *Zoological Science* **2**, 617-627.

Yanase, M., Honmura, A., Akaishi, T. & Sakuma, Y. (1988). Nerve growth factor-mediated sexual differentiation of the rat hypothalamus. *Neuroscience Research* **6**, 181-185.

Yu, W.H.A. (1982). Sex difference in the regeneration of the hypoglossal nerve in rats. *Brain Research* **238**, 404-406.

Yu. W.H.M. (1989). Administration of testosterone attenuates neuronal loss following axotomy in the brain-stem motor nuclei of female rats. *Journal of Neuroscience* **9**, 3908-3914.

Zaborszky, L. (1982). Afferent connections of the medial basal hypothalamus. *Advances in Anatomy, Embryology and Cell Biology* **69**, 1-102.

3 Factors Regulating Sexual Differentiation of the Brain: Neonatal Steroid Treatment and Development of the Reproductive Brain in the Rat

Shinji HAYASHI and Hiroaki OKAMURA

Department of Anatomy and Embryology, Tokyo Metropolitan Institute for Neurosciences, Fuchu, Tokyo 183

The nervous system of a developing animal is susceptible to environmental stimuli from both endogenous and exogenous origin. Various environmental factors were examined which affect the development of the central nervous system (CNS), including the relationship between the presence of sex steroids in perinatal rats as it influences brain functions relating to reproduction. A great body of evidence has been accumulated that administration of sex steroids into young animals such as newborn or fetus alters the morphology and function of the CNS. In this chapter, we will review the effects of sex steroids in the female rat during the neonatal period.

INDUCTION OF STERILITY BY NEONATAL ESTROGEN TREATMENT

Classical Work Explaining Sexual Differentiation of the Brain

Many early works on the rat (Takasugi, 1952, 1956; Barraclough, 1961; Barraclough & Gorski, 1961; Takewaki, 1962; Gorski, 1963; Kincl *et al.*, 1965) and in other animal species (*e.g.*, hamster: Alleva *et al.*, 1969; mouse: Nishizuka, 1974; dog: Neumann *et al.*, 1970) have shown that administration of sex steroids, either estrogen or androgen, into newborn or fetal animals evokes sterility characterized by ovulatory failure as an adult.

Since loss of a phasic change in blood gonadotropins is a common characteristic of the male animal, especially the rat, the steroid treatment to the newborn females has been deemed a production of the "male pattern", or defeminization in genetically female animals (see reviews: Doerner, 1976; Goy & McEwen, 1978; Balthazart, 1990).

Aromatization Hypothesis

Since the neuroendocrine functions are regulated mainly by the hypothalamus, on the other hand, it is conceivable that a steroid given during the neonatal period directly affects the developing hypothalamus. Naftolin and his colleagues (1972, 1975) postulated that androgen given during this period is converted to estrogen by an aromatizing enzyme available in the brain tissue and causes irreversible changes in brain function. A great body of evidence supporting this hypothesis prompted us to consider the changes caused by androgen as essentially the same as those caused by estrogen, although still with some reservations (Hayashi, 1976a).

Neonatal Steroid-sterilization and Aging

It is well known that aged female rats frequently undergo a persistent estrous state. This is accompanied by ovulatory failure and polyfollicular ovaries. In an early study, Kawashima (1960) found that uninterrupted daily injections of a very small amount of estradiol (0.01 to 0.02 μg/day) for as long as 200 days accelerated the onset of persistent estrus in the female rat. More recently, Takahashi *et al.* (1990) found the content of the luteinizing hormone-releasing hormone (LHRH) in the hypothalamus of the aged female rat to be as high as that in the neonatally estrogenized, sterile female rat which showed persistent estrus. Thus, it is conceivable that neonatal steroid treatment to the female rat is an experimental model of hypothalamic aging. Estrogen given in a large amount but only for a short period during the neonatal days accelerated the onset of sterility. This might be comparable to the effect of estrogen available in the normal control females which is secreted in a moderate amount but for a period of more than 200 days. The postulation that neonatal steroid treatment accelerates aging is supported by several authors (Naftolin *et al.*, 1990).

Steroid hormones are also known to directly affect the neuronal growth and/or morphological changes not only in perinatal animals (Raisman & Field, 1973; Torand-Allerand, 1976; Greenough *et al.*, 1977; Matsumoto & Arai, 1980, 1981, 1986; Uchibori & Kawashima, 1985) but in adults (in the rat hypothalamus: Cohen & Pfaff, 1981; Frankfurt *et al.*, 1990; and the spinal cord: Breedlove & Arnold, 1981; Kurz *et al.*, 1986; Matsumoto *et al.*, 1988; and in hamsters: Meisel & Luttrell, 1990).

SITE IN THE BRAIN WHICH IS CHANGED BY NEONATAL STEROID TREATMENT

Intracranial Application of the Steroids

Direct application of the steroids into the newborn rat brain was considered a powerful means of revealing site through which an animal's function as adult would be altered. Thus, either implantation or microinjection of the steroids was carried out intracranially in the neonatal rat. We implanted removable pellets of crystalline testosterone propionate (TP) in the hypothalamus of the neonatal female rat. After various periods of exposure, we removed the crystals and determined the minimal time required for induction of sterility. Exposure to the TP crystals for 12 hrs at the rostral hypothalamus, including the preoptic area-anterior hypothalamic area (POA-AHA) region, did not evoke sterility, while the same exposure for 24-48 hrs did cause sterility in 29-57% of those tested. Thus, minimal exposure time for induction of sterility was estimated as around 24 hrs (Hayashi & Gorski, 1974). Nadler implanted a TP-paraffin mixture in the hypothalamus of newborn female rats and obtained sterility. According to his results, the micropellets in the arcuate hypothalamic nucleus-ventromedial hypothalamic nucleus (ARC-VMH) region were more efficient in evoking sterility than those in the POA-AHA region (Nadler, 1968, 1972). However, we were unable to detect any significant difference in regional specificity in the induction of sterility throughout the hypothalamus (Hayashi & Gorski, 1974). Intrahypothalamic implants of estrogen (E) into the newborn female rats also induced sterility, but no distinct regional difference in efficacy was detected (Hayashi, 1976b).

Direct application of the steroids to the hypothalamus of neonatal rats evokes sterility, but this method is not accurate enough to determine any site-difference if the structures are located very close to each other such as the ARC, the VMH and the median eminence (ME). For example, Marcus *et al.* (1977) indicated a "running-up" of the injection vehicle containing radioactive estradiol along with the injection needle inserted into the hypothalamus. Since a diffusion of the steroids from the applied site might cause a functional and/or morphological alteration in the developing central nervous system, differences in the application site, even between the POA and the ARC, remain to be exactly determined. Thus, in order to further analyze the sites whose functions are altered by neonatal steroids, another viewpoint was needed.

Topographic Distribution of the Estrogen Receptor

Knowledge of the topographic distribution of the neurons which accumulate the steroid hormones was important to learn the exact sites which might be altered by neonatal steroid treatment. By autoradiographic (ARG) studies, Sheridan *et al.* examined the neurons which accumulate tritium labeled estradiol (^{3}H-E2) and/or testosterone (T) in both the hypothalamus and in the limbic system (Sheridan *et al.*, 1974a, b, 1975). In the hypothalamus, the E2-binding neurons were detected in the medial POA (mPO), the bed nucleus of the stria terminalis (BST), the ventrolateral part of the VMH and the ARC.

The target neurons of estrogen in the neonatal brain are also found in the amygdala, especially in the medial (AMe) and cortical (ACo) nuclei. Reports that estrogen administration to newborn rats evoked changes in the synaptic formation in the amygdala (Nishizuka & Arai, 1981a, b) suggested involvement of this structure as well as the hypothalamus in the modification of sexual activity.

The neurons which bind to tritium labeled testosterone (^{3}H-T) or its metabolites were observed in similar regions in the neonatal rat brain (Sheridan *et al.*, 1975). Thus, sites of action of neonatal E and/or T did not seem to be limited only to the rostral but also included the basal hypothalamus.

As described below, recent immunocytochemical studies using specific antibodies against estrogen receptor (ER) proteins confirmed these results of ARG studies.

LHRH IN THE HYPOTHALAMUS OF THE NEONATALLY ESTROGENIZED RAT

Rostro-caudal Distribution of the LHRH in the Hypothalamus

Immunocytochemical studies have revealed that the LHRH producing cells are located mainly in the rostral part of the forebrain such as the septum area and mPO, or in the septo-preoptic area (SPA). These neurons extend their axons mainly to the two different areas of the ventral region of the hypothalamus, *i.e.*, to the organum vasculosum lamina terminalis (OVLT) and the ME. A large amount of LHRH can be detected in these two areas (Ibata *et al.*, 1979; Merchenthaler *et al.*, 1980; Kawano & Daikoku, 1981; King *et al.*, 1982; see also Chapter 1). Although the role of LHRH stored in the OVLT is still unclear, that in the ME is well accepted. It is released into the portal vessels whose flows are directed to the anterior pituitary gland, and stimulates gonadotrophs to secrete gonadotropins (LH and FSH). The amount of LHRH in the forebrain assessed by radioimmunoassay (RIA) has a similar pattern of rostro-caudal distribution (Wheaton *et al.*, 1975; Asai & Wakabayashi, 1975).

Sterilization and Low Plasma LH

Administration of estrogen into the neonatal female rat severely suppresses reproductive activity at adult. Very small ovaries in these rats, as found in immature females indicate absence of the ovulatory surge of LH. Amount of LH in the blood is very low and administration of progesterone after estrogen priming (the steroid challenge) induces less response in the LH rise at adult than in the neonatally nontreated controls (Fig. 1) (Aihara & Hayashi, 1989; Hayashi & Aihara, 1989).

LHRH in the Hypothalamus of Sterile Rats

We measured the amount of LHRH by RIA in 400 μm thick serial frontal sections of the hypothalamic block and examined its rostro-caudal distribution in the neonatally estrogenized, sterile rat in comparison with that from the control female rat which showed regular 4-day sexual cycles. There was

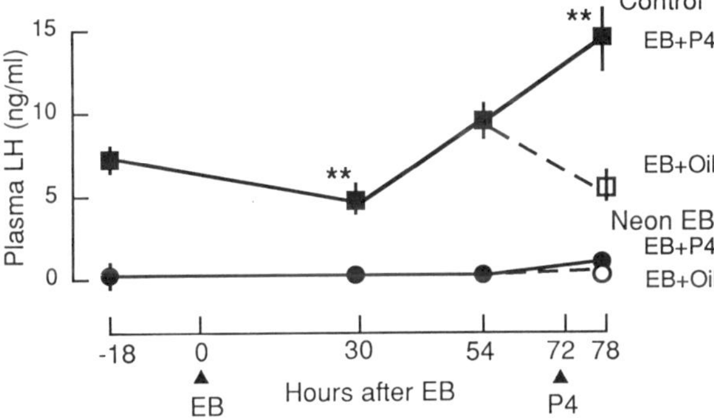

Fig. 1.   Plasma LH levels (ng/ml, expressed as NIH-LH-S1) in female rats injected with 10 μg estradiol benzoate (EB) for 1–5 days postnatally (Neon EB, circles) or neonatally nontreated controls (Control, squares). They were ovariectomized at 22 days of age. At around 50 days of age they received a priming injection of 10 μg EB per 100 g body weight at noon (0 h, indicated by EB at the bottom of the panel). Seventy-two hours later, they were injected with 2 mg progesterone or oil (indicated by P4). Blood was collected from the jugular vein under light ether anesthesia at 18 h before and 30, 54, and 78 h after the injection of EB. Open marks indicate the values with oil injection in place of P4. Vertical lines indicate the standard error of the mean (SEM). 10–12 animals were used in each group. Plasma LH levels changed between 5 and 15 ng/ml in the controls, while they remained between 0.3 and 0.5 ng/ml in the neonatally estrogenized rats. Asterisks indicate values significantly different from the preinjection level, $P < 0.01$ (Mann-Whitney U-test). (From Aihara and Hayashi, 1989).

a distinct two-peak-pattern in the distribution of LHRH; the rostral peak corresponds to the LHRH deposit in the OVLT and cell bodies in the SPA region, while the caudal peak corresponds to the LHRH storage in the ME (Fig. 2). We found that the amount of LHRH in the mediobasal hypothalamus (MBH) in the neonatally sterilized female was significantly higher than that of the control female rat (Hayashi & Aihara, 1989; Hayashi *et al.*, 1991) (Fig. 3). Higher content of LHRH in neonatally estrogenized female rats compared to control females has also been reported (Elkind-Hirsch *et al.*, 1984).

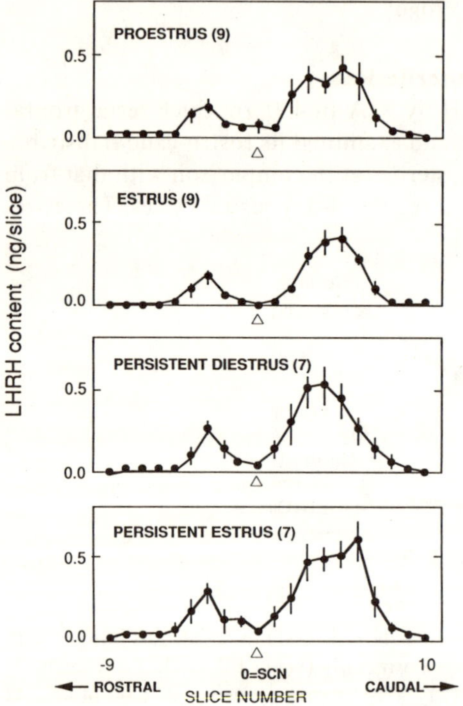

Fig. 2.   Rostro-caudal distribution of the hypothalamic LHRH levels in proestrous or estrous rats (upper two panels) and those given 10 μg EB for 5 neonatal days (lower two panels). The latter showed either persistent estrous or diestrous vaginal smears. The rat was killed at around 50 days of age and the brain was quickly frozen. The hypothalamic block was sectioned by a cryostat into frontal serial slices of 400 μm-thickness. The amount of LHRH in each slice was measured by radioimmunoassay. The frontal slice which contained the SCN was numbered "0", which is indicated by an open triangle below each panel. Thus, distance of each slice from the SCN was calculated from the serial number. Number of animals for each group is shown in the parentheses. Vertical lines indicate SEM. (From Hayashi *et al.*, 1991).

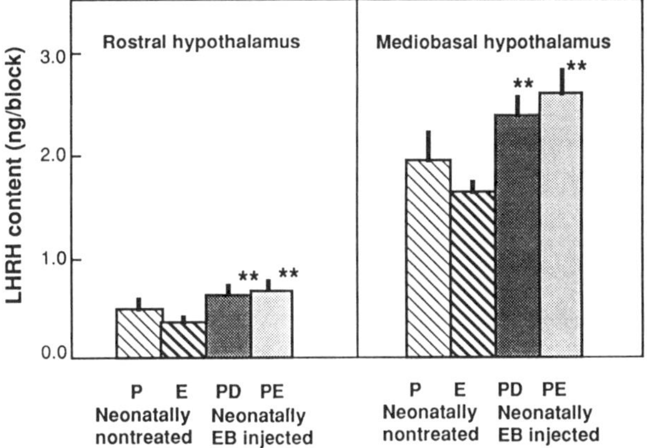

Fig. 3.   LHRH content in the rostral and mediobasal hypothalamic blocks in the rats shown in Fig. 2. Total amount of LHRH which was detected in the slices rostral to the slice with the SCN (Slice number=0 in Fig. 2) was calculated and considered the rostral block, while those from the remaining slices were considered the mediobasal block. P, E, PD and PE means proestrus, estrus, persistent diestrus and persistent estrus, respectively. Asterisks indicate significant difference from the estrus (E) control ($P<0.01$, by Newman-Keuls multiple range test). (From Hayashi *et al.*, 1991).

Ovariectomy induced a drastic decrease in LHRH in the MBH of control rats (Kelly *et al.*, 1989), while it did not induce any decrease in the LHRH in the neonatally estrogenized, sterile rats (Hayashi & Aihara, 1989; Hayashi *et al.*, 1991) (Fig. 4). This finding suggests that the neonatal estrogenization desensitizes the neuroendocrine mechanism involved in post-ovariectomy release of LHRH from the MBH, especially from the ME. Furthermore, since the amount of LHRH in the MBH of the neonatally estrogenized, sterile females was significantly higher than that of control, estrous female rats (Fig. 3, right panel), it is also suggested that the part of brain damaged neonatally with the steroid is not directly involved in the production of LHRH but in its release from the ME.

NEUROTRANSMITTERS AND NEUROPEPTIDES INVOLVED IN THE RELEASE OF LHRH

It is a rather classical postulation that brain amines are involved in the release mechanism of LHRH from the MBH (Sawyer, 1952; Krieg & Sawyer, 1976). Correlation between the sterilization by neonatal androgen and the brain noradrenergic system has also been reported (Lookingland *et al.*, 1982; Barraclough *et al.*, 1984; Handa *et al.*, 1986).

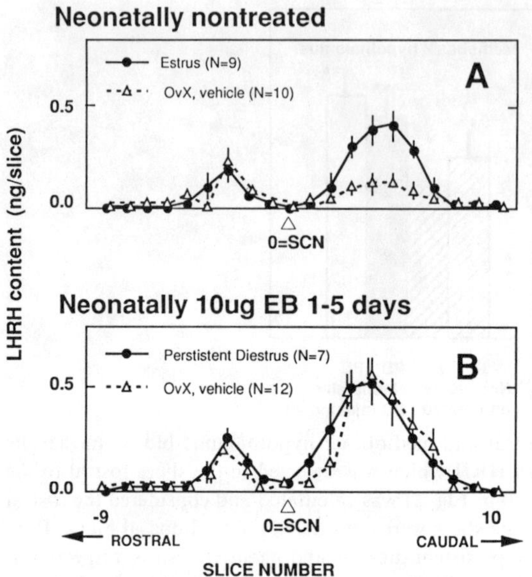

Fig. 4.   Influence of ovariectomy on the rostro-caudal distribution pattern of LHRH in the hypothalamus in neonatally nontreated (A) and neonatally estrogenized (10 μg EB for 5 neonatal days, B) female rats. Animals were killed at around 50 days of age. The control animals for panels A and B (closed circles with solid lines) are the same as in the Fig. 2, while the other rats were ovariectomized at 22 days, given oil injections at around 50 days and killed 78 h later (depicted by open triangles with broken lines in both panels). Vertical lines indicate SEM. For further details see Fig. 2. (From Hayashi & Aihara, 1989).

Neuropeptides are reportedly involved in LHRH release from the MBH. Neuropeptide Y (NPY) stimulated LHRH release from the ME when animals were given estrogen or androgen (Crowley & Kalra, 1988; Kalra *et al.*, 1988; Sahu *et al.*, 1990a, b). Stimulation of LHRH release from the ME fragment by NPY was confirmed *in vitro* by estrogen priming (Sabatino *et al.*, 1989). Conversely, NPY suppressed the pulsatile secretion of LH release from the pituitary when administered to the third ventricle in the ovariectomized female rat (McDonald *et al.*, 1989). This suppression is probably due to a decreased release of LHRH from the ME.

Involvement of substance P (SP) in gonadotropin release has been suggested by a series of morphological studies (Tsuruo *et al.*, 1984, 1987, 1991). The neuronal cell bodies which contain SP are localized in the SPA, the ventrolateral part of the VMH and the postinfundibular region of the ARC (Tsuruo *et al.*, 1991) where we also found estrogen receptor immunoreactivity, as explained in the next section. Nerve fibers and terminals which contain SP were found not only in the hypothalamus but in the amygdala (Dees & Kozlowski, 1984) where ERs are also distributed as mentioned below. SP stimulates LHRH release from the MBH *in vitro* (Ohtsuka *et al.*, 1987). Application of SP into the third ventricle also stimulates LHRH release from the MBH, whereas the intravenous injection of the same substance suppressed plasma LH (Arisawa *et al.*, 1990). When SP was injected directly into the mPO, plasma LH suppression was observed (Picanco-Diniz *et al.*, 1990).

Endorphin, in contrast, inhibits LHRH release from the ME (in the rat: Schulz *et al.*, 1981; Forman *et al.*, 1983; Kalra *et al.*, 1988; in sheep: Weesner & Malven, 1990).

Thus, participation of these neuropeptides in the LHRH release from the MBH is possibly modulated by conditions of the brain, such as the levels of amines and steroids. Modulation of the brain opioid receptors by steroids was also reported (Martini *et al.*, 1989). It is therefore possible that estrogen given during the neonatal period damages the noradrenergic and/ or peptidergic structures which are related to the release of LHRH from the ME (Almeida & Schulz, 1988; Desjardins *et al.*, 1990; Mallory & Gallo, 1990).

DISTRIBUTION OF THE ESTROGEN RECEPTOR (ER) IN THE BRAIN: RECENT STUDIES

Determination of cDNA for ER

cDNA clones of the ER from various animal species were recently isolated (human: Green *et al.*, 1986; Greene *et al.*, 1986; chicken: Krust *et al.*, 1986; mouse: White *et al.*, 1987; rat: Koike *et al.*, 1987; *Xenopus*: Weiler *et al.*, 1987; rainbow trout: Pakdel *et al.*, 1989). From these series of studies, it was revealed that the sequence of amino acids can be divided into 6 regions based on the sequence homology (A to F regions). Regions A, C and E have high degree of homology among animal species, whereas regions B, D and F are less homologous. Region C, which is the most conservative among the animal species (100% among human, rat and chicken, according to Koike *et al.*, 1987), is considered to be the DNA-binding domain, while region E is thought to be the steroid-binding domain. The N-terminal region A has a 100% homology between human and rats, while the homology was 84.4% between rat and chicken (Koike *et al.*, 1987). Thus, domain A may "play an important role in the function of the ER" (Krust *et al.*, 1986). Region D which lies between the conservative C and E is hydrophilic and considered a "hinge" between these two (Krust *et al.*, 1986).

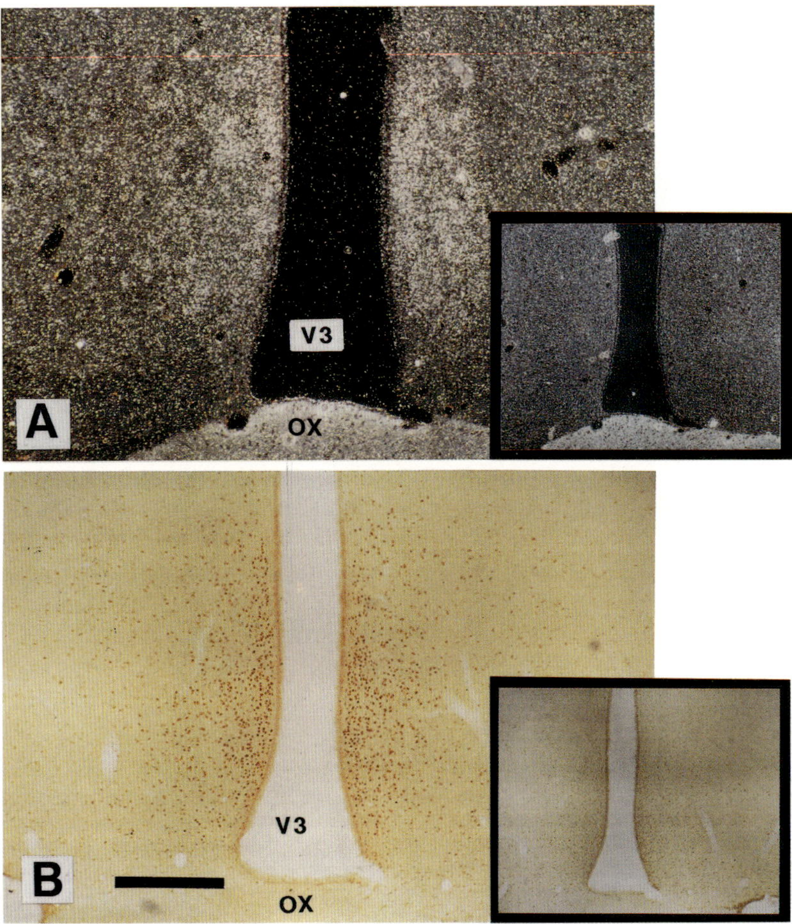

Fig. 5.   Frontal sections of the rostral part of the hypothalamus at comparable levels in the adult female rat, showing expression of mRNA for estrogen receptor (ER) (A) and the ER-like immunoreactivity (ER-LI) (B). The positive signals are found in the antero-ventral part of the medial preopic nucleus (AVPv) in both sections. Expression of the ER-mRNA was detected by *in situ* hybridization immunohistochemistry by incubation with the antisense RNA probe (A). No signals above the background levels were detected in the adjacent section incubated with the sense probe (shown in the subset). The section was counterstained with hematoxylin and observed by dark field microscopy. On the other hand, the ER-LI was detected in the cell nuclei of the AVPv in the section incubated with the anti-ER serum (B), while no ER-LI was detected in the adjacent section incubated with the preimmune serum (shown in the subset). OX, optic chiasma; V3, third ventricle. The scale indicates 0.5 mm.

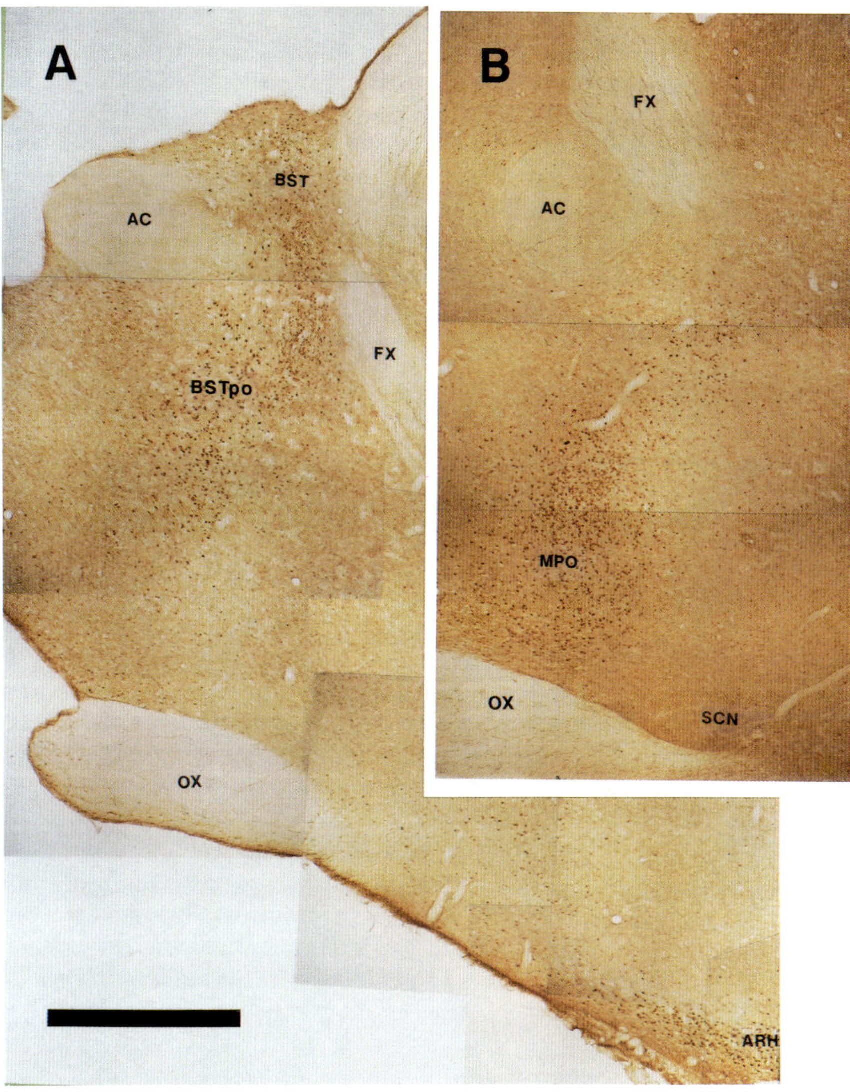

Fig. 6. Distribution of the ER-LI positive neurons in the rostral hypothalamus of the female rat. Parasagittal sections including the anterior commissure (AC), fornix (FX) and optic chiasma (OX). The parasagittal section A is located lateral to B. Groups of the immunopositive neurons are found in the medial preoptic nucleus (MPO), in the bed nucleus of the stria terminalis (BST) and in the preoptic continuation of the BST (BSTpo). In contrast, the SCN is negative for the ER-LI (B). The scale indicates 1 mm.

Detection of mRNA Coding ER
With the aid of the cDNA coding the ER, the topographical distribution of the cells which express ER-mRNA has been detected (Simerly *et al.*, 1990). Using RNA probes, we also observed expression of the ER-mRNA in the hypothalamus and the amygdala in the rat. As shown in Fig. 5A, this was found in the antero-ventral part of the mPO which is adjacent to the OVLT (AVPv, according to the definition of Simerly *et al.*, 1984). The expression was also observed in the mPO just below the anterior commissure (AC), in the ARC, in the ventrolateral part of the VMH and in the medial tuberal nucleus (MTu) in the hypothalamus; while in the amygdala it was found in the medial (AMe), the cortical (ACo) and the hippocampo-amygdala (AHi) nuclei.

Detection of ER by Antibodies
Polyclonal and monoclonal antibodies against the ER have recently been raised (Greene *et al.*, 1980a, b; Traish *et al.*, 1989; Furlow *et al.*, 1990). By application of these antibodies in immunocytochemistry, distribution of ER in the brain has been reexamined in adult female rats. The ER-like immunoreactivities (ER-LI) were revealed to be similarly located as detected by earlier ARG studies in which the radiolabelled ligand was used (Sar & Parikh, 1986; Liposits *et al.*, 1990).

We also raised a polyclonal antibody against a fusion protein of the rER and β-galactosidase, which was expressed in an *Escherichia coli* system transfected with a vector containing the full length of the rER-cDNA from a rabbit (Okamura *et al.*, in preparation). The ER-LIs were found exclusively in the cell nuclei of the following regions: In the rostral hypothalamus, the anteroventral preoptic nucleus (AVPv), the mPO, the BST, and preoptic continuation of the BST (BSTpo) (Figs. 5B & 6); in the mediobasal hypothalamus, the ARC, the ventrolateral part of the ventromedial hypothalamic nucleus (VMHvl), and the medial tuberal nucleus (MTu); in the thalamus, the lateral habenular nucleus (LHb); and in the amygdala, the medial (AMe), the baso-lateral (ABLa), the poster-olateral part of the cortical (ACo-pl) and the amygdalo-hippocampal nuclei (AHi). Topographical distribution of the ER-LI in the rostral hypothalamus is depicted in representative parasagittal sections (Fig. 6).

CONCLUSION

Administration of steroid hormones to the female rat during the late prenatal and/or the early postnatal period induces sterility. This is due to irreversible changes caused by the treatment in the neuroendocrine system involved in the reproductive functions. Since many neuroactive substances in the brain, such as steroid hormones, neurotransmitters (brain amines, amino acids, *etc.*), neuropeptides (LHRH, TRH, substance P, neuropeptide Y, *etc.*), and their receptors are related to the stimulation and/or suppression of neuroendocrine functions, analysis of the topographic distribution, amount and dynamic changes of these substances in the brain is important.

In sterile animals, a significantly higher amount of LHRH was detected in the MBH than in control animals which showed regular cyclic changes in gonadotropin secretion. Ovariectomy resulted in significant decrease in LHRH content in the hypothalamus of neonatally nontreated, control animals, while no decrease in LHRH content was noted in the neonatally estrogenized rats. Thus, neonatal estrogenization might have caused functional dysfunction not in the production of LHRH but in its releasing mechanism.

Recent development in the studies of ER allowed us to show topographical distribution of the immunoreactive ER in the rat brain. ER positive cells were distributed in the hypothalamus and amygdala in adult female rats. By *in situ* hybridization immunocytochemical method, the mRNAs coding the rat ER were detected in the hypothalamus and the amygdala both in adult and newborn females. Thus, it is probable that the neonatal treatment of the steroid hormones, at least in the case of estrogen, affects these sites and irreversibly alters gonadotropin secretion.

It is not only sex steroids, but also corticosteroids which alter neuroendocrine functions irreversibly when given during the perinatal period. Glucocorticoid delays onset of diurnal rhythmicity when administered at this time (Krieger, 1972; Miyabo & Hisada, 1975; Miyabo *et al.*, 1981; Hayashi *et al.*, 1986).

Yanase *et al.* reported a blockade of neonatal androgenization by the antibody to the nerve growth factor (Yanase *et al.*, 1988). This report supports the idea that any substance which alters the structure and function of neurons in the development contributes to the sexual differentiation of the brain, affecting the growth of neurons, dendrites, axons and synapses.

The organizational influence of steroid hormones on the CNS has been traditionally considered a specific feature of the early developmental period. However, it is now a popular belief that some structures in the brain are regulated by steroids even in adulthood. For example, dendritic spine density in the CA1 hippocampal pyramidal cells is dependent on estrogen and progesterone (Gould *et al.*, 1990). Injections of these substances also decrease the volume of the sexual dimorphic nucleus in the preoptic area, while the same procedure increases the volume of the anteroventral periventricular nucleus in gonadectomized male rats (Bloch & Gorski, 1988).

There are thus probably some common mechanisms between newborn and adult animals in the central nervous system which respond to the steroids and set plastic alterations in their structure and function.

ACKNOWLEDGEMENT

This article is dedicated to Dr. Akira Yokoyama in commemoration of his retirement from the Faculty of Agriculture of Nagoya University. One of the authors (SH) would like to express his gratitude to Dr.Yokoyama for his encouragement and valuable discussions over many years which were so very enjoyable. The authors also would like to express their thanks to Prof. Masami Muramatsu, Department of Biochemistry, Faculty of Medicine, University of Tokyo, for the generous gift of the cDNA clone for rER, and to Dr.Atsushi Kuroiwa, Tohoku University for his guidance and help in the production of the rER-β galactosidase fusion protein. Characterization of the antiserum against rER was partly done by Dr.Kazutoshi Yamamoto, Department of Biology, Waseda University. Mrs. Hiroko Ueda carried out immunocytochemistry of rER. This work was partly supported by Grants-in-Aid for Scientific Research from The Ministry of Education, Science and Culture of Japan to SH (#03640636) and to HO (#03760212).

REFERENCES

Aihara, M. & Hayashi, S. (1989). Induction of persistent diestrus followed by persistent estrus is indicative of delayed maturation of tonic gonadotropin-releasing systems. *Biology of Reproduction* **40**, 96–101.

Alleva, F.R., Alleva, J.J. & Unberger, E.J. (1969). Effects of a single prepubertal injection of testosterone propionate on later reproductive functions of the female golden hamster. *Endocrinology* **85**, 312–318.

Almeida, O.F.X. & Schulz, R. (1988). Sexual differentiation of the luteinizing hormone response of neonatal rats to the narcotic antagonist Naloxone: Critical role of estrogen receptors. *Biology of Reproduction* **39**, 1009–1012.

Arisawa, M., De Palatis, L., Snyder, G.D., Yu, W.H., Pan, G. & McCann, S.M. (1990). Stimulatory role of substance P on gonadotropin release in ovariectomized rats. *Neuroendocrinology* **51**, 523–529.

Asai, T. & Wakabayashi, K. (1975). Changes of hypothalamic LH-RF content during the rat estrous cycle. *Endocrinologia Japonica* **22**, 319–326.

Balthazart, J. (ed.) (1990). *Hormones, Brain and Behaviour in Vertebrates* (Vol. 1 & 2), Karger, Basel.

Barraclough, C.A. (1961). Production of anovulatory, sterile rats by single injections of testosterone propionate. *Endocrinology* **68**, 62–67.

Barraclough, C.A. & Gorski, R.A. (1961). Evidence that the hypothalamus is responsible for androgen-induced sterility in the female rat. *Endocrinology* **68**, 68–79.

Barraclough, C.A., Phyllis, M.W. & Selmanoff, M.K. (1984). A role for hypothalamic catecholamines in the regulation of gonadotropin secretion. *Recent Progress in Hormone Research* **40**, 487–529.

Bloch, G.J. & Gorski, R.A. (1988). Estrogen/progesterone treatment in adulthood affects the size of several components of the medial preoptic area in the male rat. *Journal of Comparative Neurology* **275**, 613–622.

Breedlove, S.M. & Arnold, A.P. (1981). Sexually dimorphic nucleus in the rat lumber spinal cord: Response to adult hormone manipulation, absence in androgen-insensitive rats. *Brain Research* **225**, 297–307.

Cohen, R.S. & Pfaff, D.W. (1981). Ultrastructure of neurons in the ventromedial nucleus of the hypothalamus in ovariectomized rats with and without estrogen treatment. *Cell and Tissue Research* **217**, 451–470.

Crowley, W.R. & Kalra, S.P. (1988). Regulation of luteinizing hormone secretion by neuropeptide Y in rats: Hypothalamic and pituitary actions. *Synapse* **2**, 276–281.

Dees, W.L. & Kozlewski, G.P. (1984). Effects of castration and ethanol on amygdaloid substance P immunoreactivity. *Neuroendocrinology* **39**, 231–235.

Doerner, G. (1976). *Hormones and Brain Differentiation*, Elsevier, Amsterdam, Oxford, London.

Desjardins, G.C., Beaudet, A. & Brawer, J.R. (1990). Alterations in opioid parameters in the hypothalamus of rats with estradiol-induced polycystic ovarian disease. *Endocrinology* **127**, 2969–2976.

Elkind-Hirsch, K., King, J.C., Gerall, A.A. & Leeman, S.E. (1984). Neonatal estrogen affects preoptic/anterior hypothalamic LHRH differently in adult male and female rats. *Neuroendocrinology* **38** , 68–74.

Forman, L.J., Sonntag, W.E. & Meites, J. (1983). Elevation of plasma LH in response to systemic injection of beta-endorphin antiserum in adult male rats. *Proceedings of Society for Experimental Biology and Medicine* **173**, 14–16.

Frankfurt, M., Gould, E., Woolley, C.S. & B.S.McEwen (1990). Gonadal steroids modify dendritic density in ventromedial hypothalamic neurons: A Golgi study in the adult rat. *Neuroendocrinology* **51**, 530–535.

Furlow, J.D., Ahrens, H., Mueller, G.C. & Gorski, J. (1990). Antisera to a synthetic peptide recognize native and denatured rat estrogen receptors. *Endocrinology* **127**, 1028–1032.

Gorski, R.A. (1963). Modification of ovulatory mechanisms by postnatal administration of estrogen in female rats. *American Journal of Physiology* **205**, 842–844.

Gould, E., Woolley, C.S., Frankfurt, M. & McEwen, B. (1990). Gonadal steroids regulate dendritic spine density in hippocampal pyramidal cells in adulthood. *Journal of Neuroscience* **10**, 1286–1291.

Goy, R.W. & McEwen, B.S. (1978). *Sexual Differentiation of the Brain*, MIT Press, Cambridge, Massachusetts and London.

Green, S., Walter, P., Kumar, V., Krust, A., Bornert, J.-M., Argos, P. & Chambon, P. (1986). Human estrogen receptor cDNA: Sequence, expression, and homology to v-*erb*-A. *Nature* **320**, 134–139.

Greene, G.L., Fitch, F.W. & Jensen, E.V. (1980a). Monoclonal antibodies to estrophilin: Probes for the study of estrogen receptors. *Proceedings of National Academy of Sciences, U.S.A.* **77**, 157–161.

Greene, G.L., Gilna, P., Waterfield, M., Baker, A., Hort, Y. & Shine, J. (1986). Sequence and expression of human estrogen receptor complementary DNA. *Science* **231**, 1150–1154.

Greene, G.L., Nolan, C., Engler, J.P. & Jensen, E.V. (1980b). Monoclonal antibodies to human estrogen receptor. *Proceedings of National Academy of Sciences, U.S.A.* **77**, 5115-5119.

Greenough, W.T., Carter, C.S., Steerman, C. & DeVoogt, T.J. (1977). Sex differences in dendritic patterns in hamster preoptic area. *Brain Research* **126**, 63-72.

Handa, R.J., Condon, T.P., Whitmoyer, D.I. & Gorski, R.A. (1986). Gonadotropin secretion following intraventricular norepinephrine infusion into neonatally androgenized female rats. *Neuroendocrinology* **43**, 269-272.

Hayashi, S. (1976a). Failure of intrahypothalamic implants of an estrogen antagonist, ethamoxytriphetol (MER-25), to block neonatal androgen-sterilization. *Proceedings of Society of Experimental Biology and Medicine* **152**, 389-392.

Hayashi, S. (1976b). Sterilization of female rats by neonatal placement of estradiol micropellets in anterior hypothalamus. *Endocrinologia Japonica* **23**, 55-60.

Hayashi, S. & Aihara, M. (1989). Neonatal estrogenization of the female rat: A useful model for analysis of hypothalamic involvement of gonadotropin release. *Zoological Science* **6**, 1059-1068.

Hayashi, S., Aihara, M & Wakabayashi, K. (1991). Content and distribution pattern of luteinizing hormone-releasing hormone (LHRH) in the hypothalamus of neonatally estrogenized rats. *Neuroscience Research* **12**, 366-378.

Hayashi, S. & Gorski, R.A. (1974). Critical exposure time for androgenization by intracranial crystals of testosterone propionate. *Endocrinology* **94**, 1161-1167.

Hayashi, S., Ooya, E. & Miyabo, S. (1986). Effects of intracranial implantation of ovarian and adrenal steroids and hypothalamic deafferentation in newborn female rats on onset of biological rhythmicity. *Monograph of Neural Sciences* **12**, 58-63.

Ibata, Y.K., Watanabe, H., Kinoshita, H. Kubo, S. & Sano, Y. (1979). The location of LH-RH neurons in the rat hypothalamus and their pathways to the median eminence. *Cell and Tissue Research* **198**, 381-395.

Kalra, S.P., Allen, L.G., Sahu, A., Kalra, P.S. & Crowley, W.R. (1988). Gonadal steroid and neuropeptide Y-opioid-LHRH axis: interactions and diversities. *Journal of Steroid Biochemistry* **30**, 185-193.

Kawano, H. & Daikoku, S. (1981). Immunohistochemical demonstration of LHRH neurons and their pathways in the rat hypothalamus. *Neuroendocrinology* **32**, 179-186.

Kawashima, S. (1960). Influence of continued injections of sex steroids on the estrous cycle in the adult rat. *Annotationes Zoologicae Japonenses* **33**, 226-233.

Kelly, M.J., Garrett, J., Bosch, M.A., Roselli, C.E., Douglass, J., Adelman, J.P. & Ronnekleiv, O.K. (1989). Effects of ovariectomy on GnRH mRNA, proGnRH and GnRH levels in the preoptic hypothalamus of the female rat. *Neuroendocrinology* **49**, 88-97.

Kincl, F.A., Folch Pi, A., Maqueo, M., Herrera Lasso, L., Oriol, A. & Dorfman, R.I. (1965). Inhibition of sexual development in male and female rats treated with various steroids at the age of five days. *Acta Endocinologica (Kbh)* **49**, 193-206.

King, J.C., Tobet, S.A., Snavely, F.L. & Arimura, A. (1982). LHRH immunopositive cells and their projections to the median eminence and organum vasculosum of the lamina terminalis. *Journal of Comparative Neurology* **209**, 287-300.

Koike, S., Sakai, M. & Muramatsu, M. (1987). Molecular cloning and characterization of rat estrogen receptor cDNA. *Nucleic Acids Research* **15**, 2499-2513.

Krieg, R.J. & Sawyer, C.H. (1976). Effects of intraventricular catecholamines on luteinizing hormone release in ovariectomized steroid primed rats. *Endocrinology* **99**, 411-419.

Krieger, D.T. (1972). Circadian corticosteroid periodicity: Critical period for abolition by neonatal injection of corticosteroid. *Science* **78**, 1205-1207.

Krust, A., Green, S., Argos, P., Kumar, V., Walter, P., Bornet, J.-M. & Chambon, P. (1986). The chicken oestrogen receptor sequence: homology with v-*erb*A and the human oestrogen glucocorticoid receptors. *EMBO Journal* **5**, 891-897.

Kurz, E.M., Sengelaub, D.R. & Arnold, A.P. (1986). Androgens regulate the dendritic length of mammalian motoneurons in adulthood. *Science* **232**, 395-398.

Liposits, Z., Kallo, I., Coen, C.W., Paull, W.K. & Flerko, B. (1990). Ultrastructural analysis of estrogen receptor immunoreactive neurons in the medial preoptic area of the female rat brain. *Histochemistry* **93**, 233-239.

Lookingland, K., Wise, P.M. & Barraclough, C.A. (1982). Failure of the hypothalamic noradrenergic system to function in adult androgen sterilized rats. *Biology of Reproduction* **27**, 268-281.

Mallory, D.S. & Gallo, R.V. (1990). Medial preoptic-anterior hypothalamic area involvement in the suppression of pulsatile LH release by a mu-opioid agonist in the ovariectomized rat. *Brain Research Bulletin* **25**, 251-257.

Marcus, D.S., Schuler, H., Boccella, L., Zivic, W. & Josimovich, J.B. (1977). New technique of injection of estradiol into preoptic-anterior hypothalamic area of newborn rats; technique limiting diffusion and duration, with preliminary results on postpuberal block of ovulation. *Endocrinology* **100**, 862-872.

Martini, L., Dondi, D., Limonta, P., Maggi, R. & Piva, F. (1989). Modulation by sex steroids of brain opioid receptors: Implications for the control of gonadotropins and prolactin secretion. *Journal of Steroid Biochemistry* **33**, 673-681.

Matsumoto, A. & Arai, Y. (1980). Sexual dimorphism in "wiring pattern" in the hypothalamic arcuate nucleus and its modification by neonatal hormonal environment. *Brain Research* **190**, 238-242.

Matsumoto, A. & Arai, Y. (1981). Effects of androgen on sexual differentiation of synaptic organization in the hypothalamic arcuate nucleus: An ontogenetic study. *Neuroendocrinology* **33**, 166-169.

Matsumoto, A. & Arai, Y. (1986). Male-female difference in synaptic organization of the ventromedial nucleus of the hypothalamus in the rat. *Neuroendocrinology* **42**, 232-236.

Matsumoto, A., Arnold, A.P., Zampighi, G.A. & Micevych, P.E. (1988). Androgenic regulation of gap junctions between motoneurons in rat spinal cord. *Journal of Neuroscience* **8**, 4177-4183.

McDonald, J.K., Lumpkin, M.D. & Depaolo, L.V. (1989). Neuropeptide Y suppresses pulsatile secretion of luteinizing hormone in ovariectomized rats: Possible site of action. *Endocrinology* **125**, 186-191.

Meisel, R.L. & Luttrell, V.R. (1990). Estradiol increases the dendritic length of ventromedial hypothalamic neurons in female Syrian hamsters. *Brain Research Bulletin* **25**, 165-168.

Merchenthaler, I., Kovács, G., Lovász, G. & Sétáló, G. (1980). The preoptico-infundibular LH-RH tract of the rat. *Brain Research* **198**, 63-74.

Miyabo, S. & Hisada, T. (1975). Sex difference in ontogenesis of circadian adrenocortical rhythm in cortisone-primed rats. *Nature (London)* **256**, 590-592.

Miyabo, S., Ooya, E. & Hayashi, S. (1981). Effect of intrahypothalamic implantation of cortisone acetate on the onset of circadian corticosterone rhythm in neonatal female rats. *Neuroendocrinology* **33**, 47-51.

Nadler, R.D. (1968). Masculinization of female rats by intracranial implantation of androgen in infancy. *Journal of Comparative Physiology and Psychology* **66**, 157–167.

Nadler, R.D. (1972). Intrahypothalamic locus for induction of androgen sterilization in neonatal female rats. *Neuroendocrinology* **9**, 349–357.

Naftolin, F., Garcia-Segura, L.M., Keefe, D., Leranth, C., MacLusky N, J. & Brawer, J.R. (1990). Estrogen effects on the synaptology and neural membranes of the rat hypothalamic arcuate nucleus. *Biology of Reproduction* **42**, 21–28.

Naftolin, F., Ryan, K.J. & Petro, Z. (1972). Aromatization of androstenedione by the anterior hypothalamus of adult male and female rats. *Endocrinology* **90**, 295–298.

Naftolin, F., Ryan, K.J., Davies, I.J., Reddy, V.V., Flores, F., Petro, Z., Kuhn, M., White, R.J., Takaoka, Y. & Wolin L. (1975). The formation of estrogens by central neuroendocrine tissues. *Recent Progress in Hormone Research* **31**, 295–316.

Neumann, F., von Berswordt-Wallrabe, R., Elger, W., Steinbeck, H., Hahn, J.D. & Kramer, M. (1970). Aspects of androgen-dependent event as studied by antiandrogens. *Recent Progress in Hormone Research* **26**, 337–410.

Nishizuka, M. (1974). Induction of persistent estrus by a single neonatal injection of steroids in the mouse. *Journal of Faculty of Science, University of Tokyo, Section IV* **13**, 149–157.

Nishizuka, M. & Arai, Y. (1981a). Sexual dimorphism in synaptic organization in the amygdala and its dependence on neonatal hormone environment. *Brain Research* **212**, 31–38.

Nishizuka, M. & Arai, Y. (1981b). Organizational action of estrogen on synaptic pattern in the amygdala: Implications for sexual differentiation of the brain. *Brain Research* **213**, 422–426.

Ohtsuka, S., Miyake, A., Nishizaki, T., Tasaka, K., Aono, T. & Tanizawa, O. (1987). Substance P stimulates gonadotropin-releasing hormone release from rat hypothalamus *in vitro* with involvement of oestrogen. *Acta Endocrinologica (Kbh)* **115**, 247–252.

Pakdel, F., Le Guellec, C., Vaillant, C., Le Roux, M.G. & Valotaire, Y. (1989). Identification and estrogen induction of two estrogen receptors (ER) messenger ribonucleic acids in the rainbow trout liver: Sequence homology with other ERs. *Molecular Endocrinology* **3**, 44–51.

Picanco-Diniz, D.L., Valenca, M.M., Franci, C.R. & Antunes-Rodrigues, J. (1990). Role of substance P in the medial preoptic area in the regulation of gonadotropin and prolactin secretion in normal and orchidectomized rats. *Neuroendocrinology* **51**, 675–682.

Raisman, G. & Field, P.M. (1973). Sexual dimorphism in the neuropile of the prooptic area of the rat and its dependence on neonatal androgen. *Brain Research* **54**, 1–29.

Sabatino, F.D., Collins, P. & McDonald, J.K. (1989). Neuropeptide-Y stimulation of luteinizing hormone-releasing hormone secretion from the median eminence *in vitro* by estrogen-dependent and extracellular Ca^{2+}-independent mechanisms. *Endocrinology* **124**, 2089–2098.

Sahu, A., Crowley, W.R. & Kalra, S.P. (1990a). An opioid-neuropeptide-Y transmission line to luteinizing hormone (LH)-releasing hormone neurons: A role in the induction of LH surge. *Endocrinology* **126**, 876–883.

Sahu, A., Kalra, P.S., Crowley, W.R. & Kalra, S.P. (1990b). Functional heterogeneity in neuropeptide-Y-producing cells in the rat brain as revealed by testosterone action. *Endocrinology,* **127**, 2307–2312.

Sar, M. & Parikh, I. (1986). Immunohistochemical localization of estrogen receptor in rat brain, pituitary and uterus with monoclonal antibodies. *Journal of Steroid Biochemistry* **24**, 497–503.

Sawyer, C.H. (1952). Stimulation of ovulation in the rabbit by intraventricular injection of epinephrine or norepinephrine. *Anatomical Record* **112**, 385.

Schulz, R., Wilhelm, A., Pirke, K.M., Gramsch, C. & Herz, A. (1981). Beta-endorphin and dynorphin control serum luteinizing hormone level in immature female rats. *Nature* **294**, 757–759.

Sheridan, P.J., Sar, M. & Stumpf, W.E. (1974a). Autoradiographic localization of ^{3}H-estradiol or its metabolites in the central nervous system of the developing rat. *Endocrinology* **94**, 1386–1390.

Sheridan, P.J., Sar, M. & Stumpf, W.E. (1974b). Interactions of exogenous steroids in the developing rat brain. *Endocrinology* **95**, 1749–1754.

Sheridan, P.J., Sar, M. & Stumpf, W.E. (1975). Estrogen and androgen distribution in the brain of neonatal rats. In: *Anatomical Neuroendocrinology* (Stumpf, W.E. and Grant, L.D., eds.), Karger, Basel, pp. 134–141.

Simerly, R.B., Chang, C., Muramatsu, M. & Swanson, L. (1990). Distribution of androgen and estrogen receptor mRNA-containing cells in the rat brain: An *in situ* hybridization study. *Journal of Comparative Neurology* **294**, 76–95.

Simerly, R.B., Swanson, L.W. & Gorski, R.A. (1984). Demonstration of a sexual dimorphism in the distribution of serotonin-immunoreactive fibers in the medial preoptic nucleus of the rat. *Journal of Comparative Neurology* **225**, 151–166.

Takahashi, S., Nomura, K. & Kawashima, S. (1990). Hypothalamic luteinizing hormone-releasing hormone (LHRH) contents and LH secretion in aging female rats. *Endocrinologia Japonica* **37**, 213–221.

Takasugi, N. (1952). Einfluesse von Androgen und Estrogen auf die Ovarien der neugeborenen und reifen, weiblichen Ratten. *Annotationes Zoologicae Japonenses* **25**, 120–127.

Takasugi, N. (1956). Veraenderungen der hypophysaeren, gonadotropen Aktivitaet der daueranoestrischen und der ovariektomierten Ratten unter Stress-Situationen. *Journal of Faculty of Science, University of Tokyo, Section IV* **7**, 625–639.

Takewaki, K. (1962). Some aspects of hormonal mechanism involved in persistent estrus in the rat. *Experientia* **18**, 1–6.

Torand-Allerand, C.D. (1976). Sex steroids and the development of the newborn mouse hypothalamus and preoptic area *in vitro*: Implications for sexual differentiation. *Brain Research* **106**, 407–412.

Traish, A., Kim, N. & Wotiz, H. (1989). Characterization of polyclonal antibodies to preselected domains of the human estrogen receptor. *Endocrinology* **125**, 172–179.

Tsuruo, Y., Hisano, S., Nakanishi, J., Katoh, S. & Daikoku, S. (1987). Immunohistochemical studies on the roles of substance P in the rat hypothalamus: Possible implication in the hypothalamic-hypophysial-gonadal system. *Neuroendocrinology* **45**, 389–401.

Tsuruo, Y., Hisano, S., Okamura, Y., Tsukamoto, N. & Daikoku, S. (1984). Hypothalamic substance P-containing neurons. Sex-dependent topographical differences and ultrastructural transformations associated with stages of the estrous cycle. *Brain Research* **305**, 331–341.

Tsuruo, Y., Kawano, H., Hisano, S., Kagotani, Y., Daikoku, S., Zhang, T. & Yanaihara, N. (1991). Substance P-containing neurons innervating LHRH-containing neurons in the septo-preoptic area of rats. *Neuroendocrinology* **53**, 236–245.

Uchibori, M. & Kawashima, S. (1985). Effects of sex steroids on the growth of neuronal processes in neonatal rat hypothalamus-preoptic area and cerebral cortex in primary culture. *International Journal of Developmental Neuroscience* **3**, 169–176.

Wheaton, J.E., Krulich, L. & McCann, S.M. (1975). Localization of luteinizing hormone-releasing hormone in the preoptic area and hypothalamus of the rat using radioimmunoassay. *Endocrinology* **97**, 30–38.

Weesner, G.D. & Malven, P.V. (1990). Intracerebral immunoneutralization of beta-endorphin and meta-enkephalin disinhibits release of pituitary luteinizing hormone in sheep. *Neuroendocrinology* **52**, 382–388.

White, R., Lees, J.A., Needham, M. & Parker, M. (1987). Structural organization and expression of the mouse estrogen receptor. *Molecular Endocrinology* **1**, 735–744.

Weiler, I., Lew, D. & Shapiro, D. (1987). The *Xenopus laevis* estrogen receptor. Sequence homology with human and avian receptors and identification of multiple estrogen receptor messenger ribonucleic acids. *Molecular Endocrinology* **1**, 355–362.

Yanase, M., Honmura, A., Akaishi, T. & Sakuma, Y. (1988). Nerve growth factor-mediated sexual differentiation of the rat hypothalamus. *Neuroscience Research* **6**, 181–185.

4 Generating Mechanism of Pulsatile Gonadotropin Release

Masugi NISHIHARA, Hiromi HIRUMA* and Fukuko KIMURA*

*Department of Veterinary Physiology, Veterinary Medical Science, The University of Tokyo, Tokyo 113 and *Department of Physiology, Yokohama City University School of Medicine, Yokohama 232*

Gonadal function in mammals depends on gonadotropins secreted from the pituitary gland in a pulsatile manner (Knobil, 1980; Lincoln, 1988). The pulsatile pattern of gonadotropin secretion was first found in the ovariectomized monkey (Dierschke *et al.*, 1970) and then extended to either intact or castrated animals in many species of both sexes. Each species of animals has its own specific pulse pattern characteristics. For example, the interval of luteinizing hormone (LH) pulses in the ovariectomized monkey, goat and rat is about 60, 40 and 20 min, respectively (Wilson *et al.*, 1984; Mori *et al.*, 1991; Kimura *et al.*, 1991). Deviation of the pulse frequency from the physiological one results in the regression of gonadal function, in disturbances of the reproductive process and in infertility (Lincoln & Short, 1980; Pohl *et al.*, 1983).

The pulsatile secretion of gonadotropin is elicited by the periodic secretion of LH-releasing hormone (LHRH) into the pituitary portal circulation from the hypothalamus (Clarke & Cummins, 1982). This pulsatility of LHRH secretion is essential for the maintenance of normal gonadotropin secretion. It has been shown that the continuous infusion of LHRH or intermittent administration of LHRH at supraphysiological frequencies leads to complete refractoriness of the pituitary gland with a resultant cessation of gonadotropin secretion (Wildt *et al.*, 1981). Conversely, a decrease in the frequency of pulsatile LHRH secretion, which is seen in, for example, stressed animals or seasonal breeders in nonbreeding seasons, is incompatible with normal gonadal function though each LHRH pulse can be followed by a gonadotropin pulse.

The pulsatility of LHRH secretion is governed, in turn, by the periodic activation of a neuronal system, a pulse generator, in the medial basal hypothalamus (Wilson *et al.*, 1984). Thus, many internal and external environmental factors including sex steroids, endogenous opioid peptides, corticotropin-releasing factor (CRF), stress, suckling, photoperiod and feeding conditions first modify the electrical activity of the central LHRH pulse generator, which then affect the pulsatile pattern of gonadotropin secretion and, finally, the reproductive function. These factors act on the LHRH pulse generator *via* afferent connections that utilize a variety of monoamines, amino acids, and peptides for neurotransmission (Lincoln, 1988).

The mechanisms of modifying the LHRH pulse generator activity have largely been deduced from changes in LH profiles in peripheral blood after experimental manipulations, since sampling of the pituitary portal blood remains very difficult. To evaluate blood LH profiles, however, many additional factors, such as the direct effects of experimental manipulations on the pituitary, the sensitivity of the pituitary to LHRH, and the method for LH pulse detection have to be taken into account. Thus, for the study of the central regulation of the reproductive function, a new index which reflects the activity of the LHRH pulse generator with greater precision has been sought.

Characteristic increases in the neuronal activity, each of which is associated with the initiation of pulsatile LH secretion, have been recorded in the medial basal hypothalamus of the monkey, rat and goat by means of multiunit activity (MUA) recording techniques (Kawakami *et al.*, 1982; Wilson *et al.*, 1984; Kimura *et al.*, 1991; Nishihara *et al.*, 1991; Mori *et al.*, 1991). An unambiguous unitary relationship between the increased electrical activity (volley) and the LH pulse under a variety of physiological and experimental conditions indicates that the MUA volleys represent the electrical activity of the LHRH pulse generator. Using the MUA as an index, it is now possible to access directly the hypothalamic controlling system which governs the pulsatile secretion of LHRH.

In this chapter, the characteristics of the hypothalamic electrical activity associated with pulsatile LH secretion in three mammalian species, the monkey, rat and goat, and recent studies on the neuroendocrine control of the LHRH pulse generator activity are described.

ELECTRICAL ACTIVITY OF THE LHRH PULSE GENERATOR

The recording system for the electrical activity of the hypothalamic LHRH pulse generator from conscious animals was first developed in the monkey (Wilson *et al.*, 1984). Because the release of sufficient LHRH to elicit an LH pulse undoubtedly requires the synchronous discharge of LHRH neurons, MUA recording techniques using chronically implanted multiple electrode arrays that blanketed the entire medial basal hypothalamus were employed to detect changes in the neuronal activity associated with changes in blood LH level. Further, to obtain very stable recordings, a buffer amplifier with high input impedance and low output impedance was directly connected to the electrode assembly and the differential recording between two electrodes was carried out. By means of similar MUA recording techniques, electrophysiological correlates of the pulsatile LH secretion have recently been described in the free moving rat (Kimura *et al.*, 1991; Nishihara *et al.*, 1991) and goat (Mori *et al.*, 1991). Thus, this electrical activity seems to be common among mammals and can be a good index for the reproductive function.

An example of the rate meter records of hypothalamic MUA and serum LH profiles in an ovariectomized rat is shown in Fig. 1. Each MUA volley is associated with or slightly precedes an LH pulse and is characterized by an initial brief overshoot followed by a gradual decrease to the baseline activity. The ovariectomized monkey, rat and goat have their

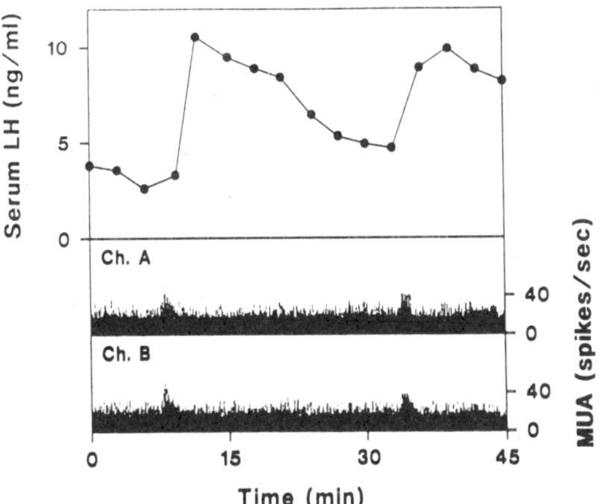

Fig. 1.   Correlation between LH pulses and MUA volleys in a freely moving ovariectomized rat. Blood samples were taken every 3 min. In this animal, MUA volleys were recorded from 2 different electrodes. (From Kimura *et al.*, 1991).

specific MUA volley characteristics. For example, the interval between the volleys is about 60, 40 and 20 min in the ovariectomized monkey, goat and rat, respectively, and volley duration is about 15, 4 and 2 min. The MUA volleys in the ovariectomized monkey have a plateau phase, which is not as obvious in the ovariectomized rat and goat, following the initial overshoot (Wilson *et al.*, 1984).

Interestingly, the intact female monkey during the follicular phase of the menstrual cycle has MUA volleys much shorter in duration (1–3 min) and devoid of plateau phase (Williams *et al.*, 1988). After ovariectomy, volley duration progressively increases for a few months. This increase in duration may reflect the increase in the excitability of the LHRH pulse generator due to the removal of an inhibitory effect of sex steroids on it, since estradiol administered at physiological levels shortens volley duration in ovariectomized monkeys (Kesner *et al.*, 1987). It has been further shown that estrogen affects the excitability of hypothalamic arcuate neurons directly (Nishihara & Kimura, 1989) or indirectly by changing the sensitivity to neurotransmitters (Nishihara *et al.*, 1986). Moreover, morphological changes such as synaptic remodelling caused by estrogen (Garcia-Segura *et al.*, 1986, see also Chapter 2) may also be involved in gradual changes in volley duration.

In contrast to the duration, the frequency of MUA volleys following ovariectomy is almost the same as that during the follicular phase, allowing speculation that the duration and frequency of these volleys are regulated independently. Observations that both barbiturates and morphine reduce volley frequency, while barbiturates lengthen and morphine reduces volley duration (Wilson *et al.*, 1984; Williams *et al.*, 1990b) support this speculation. The mechanisms regulating the duration and frequency of MUA volleys, however, remain to be resolved. All that can be said at present is that short-loop and ultrashort-loop feedback actions of LH and LHRH on the LHRH pulse generator, respectively, do not seem to be involved in the generator's rhythmic activation and deactivation, because systemic (Kesner *et al.*, 1986a) or intracerebroventricular administrations of LHRH, its agonist or antagonist have an effect on neither volley duration nor frequency.

The frequency of MUA volleys directly determines frequency of LHRH pulses, and the role of changing frequency in the control of gonadotropin secretion has been clearly demonstrated as mentioned. Although the physiological significance of changes in the duration of MUA volleys is currently unclear, the relationship between the volley duration and dynamics of LH pulses has been extensively studied in the ovariectomized monkey (Williams *et al.*, 1990c). Since no significant correlation has been reported between volley duration and the magnitude of LH pulses, all of the LHRH secreted per pulse seems to be released at the onset of each MUA volley. The remainder of the increase in electrical activity is unlikely to have further influence on LHRH secretion, although effects on other systems cannot be excluded.

By recording the MUA, continuous monitoring of the LHRH pulse generator activity has been possible, and the diurnal variation in the frequency of MUA volleys has been examined in the monkey, rat and goat. In the monkey, as anticipated from previous studies of LH pulse frequency, the frequency of MUA volleys is markedly reduced at night (O'Byrne *et al.*, 1990). In the rat, however, this frequency, *i.e.*, the frequency of LH pulses is increased during the dark phase (Hiruma *et al.*, 1992). The volley frequency in the goat does not change throughout the day. These findings suggest that the circadian oscillation of the LHRH pulse generator activity varies from species to species.

The MUA volleys associated with LH pulses were recorded from the arcuate nucleus-median eminence regions of the hypothalamus in all three species of animals studied. At present, the precise cellular nature, the activity of which was recorded, is still not clear. However, the histological finding in the rat that the tip of most of the active electrodes, from which the MUA volleys were recorded, was located in the median eminence may indicate that the volleys are recorded from the axonal tract or terminal field of LHRH neurons. In the rat, no MUA volleys associated with LH pulses were recorded in the preoptic area, although the majority of LHRH neurons are located in this area (Kawano & Daikoku, 1981). The reason may be that LHRH neurons are very few in number and widely scattered even in the preoptic area (Wray & Hoffman, 1986). Alternatively, the preoptic area is not substantially involved in generating pulsatile LH secretion in ovariectomized rats, since LH pulses are maintained even after the complete deafferentation of the hypothalamus (Blake & Sawyer, 1974; Soper & Weick, 1980; Ohkura *et al.*, 1991). The possibility that the MUA volleys are not discharged from LHRH neurons but from a neuronal oscillator residing in the medial basal hypothalamus which then activates LHRH neurons can not be ruled out.

MUA recording techniques have enabled the chronic monitoring of the electrical activity of the hypothalamic LHRH pulse generator in real time from freely moving animals. This is an advantage, especially for rats from which frequent blood sampling over a long period is very difficult. Further, experimental treatments can be applied at known intervals from the onset of MUA volley, *i.e.*, the initiation of LH pulse. Although many drugs and hormones used for neuroendocrinological study act on both the pituitary and hypothalamic levels, MUA monitoring can be used to evaluate their direct central action. The MUA volleys can be a simple and good index for the effects of stress, pheromones, feeding conditions, *etc.*, on the reproductive function, since these volleys are very sensitive to factors affecting gonadotropin secretion and are easy to evaluate (Williams *et al.*, 1990a, 1990b).

It has been suggested that many putative neurotransmitters including catecholamines (Ramirez *et al.*, 1984), excitatory (Gay & Plant, 1987) and inhibitory (Lamberts *et al.*, 1983; Nishihara & Kimura, 1987b) amino acids and peptides (Ferin *et al.*, 1984, Kaynard *et al.*, 1990), and hormones (Knobil, 1980; Lincoln, 1988) can modulate the pulsatile secretion of gonadotropins. Experiments are now being performed to elucidate how neurotransmitters and hormones are involved in the control of the LHRH pulse generator activity by investigating changes in both MUA volleys and LH pulses induced by the administration of hormones, agonists and antagonists of neurotransmitters. Our recent studies on the opioid peptidergic and noradrenergic systems in this regard are described below.

OPIOID PEPTIDERGIC SYSTEM AND THE LHRH PULSE GENERATOR

There is a general agreement that opioids have an inhibitory effect on gonadotropin secretion. Stimulation of opioid receptors by morphine or an endogenous opioid peptide β-endorphin has been shown to suppress preovulatory LH surge in rats (Pang *et al.*, 1977; Kubo *et al.*, 1983) as well as pulsatile LH secretion in rats and primates (Kinoshita *et al.*, 1980; Ferin *et al.*, 1982). On the other hand, administration of antibodies to β-endorphin and dynorphin (Schulz *et al.*, 1981) or an opioid receptor antagonist, naloxone (Piva *et al.*, 1985; Allen & Kalra, 1986), stimulates LH secretion. Further, it has been clearly demonstrated that opioids have an effect on LH secretion by acting on the central nervous system (Kesner *et al.*, 1986b).

The stimulatory effect of naloxone on gonadotropin secretion has been preferentially described in intact, but not in gonadectomized animals. A notion that endogenous opioid peptides mediate an inhibitory effect of gonadal steroids on gonadotropin secretion has been proposed (Bhanot & Wilkinson, 1983; Petraglia *et al.*, 1984). For example, it has been suggested that progesterone reduces the frequency of LH pulses *via* endogenous opioid peptide-mediated mechanisms in both the monkey (Ferin *et al.*, 1984) and the rat (Devorshak-Harvey *et al.*, 1987). In the monkey, naloxone has been shown to have no effect on LH pulses after gonadectomy (Kesner *et al.*, 1986b). However, in the rat, it is still not clear whether endogenous opioid peptides affect LH secretion in the absence of gonadal steroids. Thus, the involvement of endogenous opioid peptides in the control of the LH pulse generator activity was studied in the ovariectomized rat using MUA as an index (Kimura *et al.*, 1991).

Naloxone was intravenously infused at doses of 0.1, 0.2 and 0.5 mg/kg/h for 1 h into the ovariectomized rats showing MUA volleys associated with LH pulses. As shown in Fig. 2, naloxone increased volley frequency

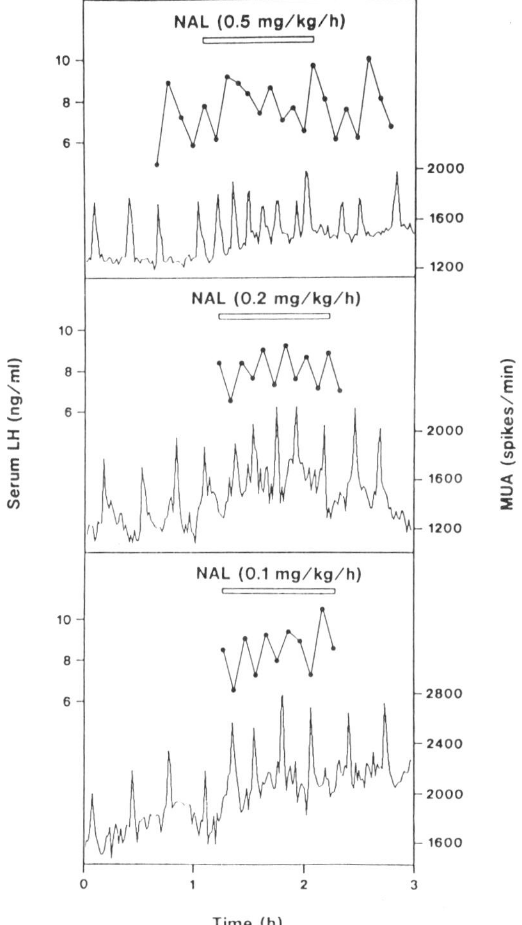

Fig. 2. Effects of intravenous infusion (indicated by horizontal bars) of naloxone (NAL) on LH pulses and MUA volleys in a freely moving ovariectomized rat. Blood samples were taken every 6 min. (From Kimura *et al.*, 1991.)

in a dose dependent manner. The frequency during the infusion of 0.5 mg/ kg/h naloxone was about 2 times higher than that during the pretreatment control period. When naloxone was infused at a dose of less than 0.2 mg/ kg/h, each MUA volley was followed by a clear LH pulse in this experiment. During the infusion of naloxone of 0.5 mg/kg/h, however, the clear pulsatility of LH secretion observed in the pretreatment control period disappeared, suggesting that the frequency of LHRH pulses was too high for the pituitary to secrete pulsatile LH in response to each LHRH pulse. Alternatively, the interval of blood sampling (6 min) was too long to detect each LH pulse. It has been reported in both *in vivo* and *in vitro* examinations that an increase in the frequency of the pulsatile administration of LHRH resulted in an increase in the mean LH concentration and a decrease in the amount of LH secreted in response to each LH pulse if the frequency remained within the physiological range (Clarke *et al.*, 1984; Kamel *et al.*, 1987).

The effects of naloxone infusion on the duration and the interval of the MUA volleys are summarized in Fig. 3. Naloxone increased the duration and decreased the interval of the MUA volleys in a dose dependent manner. Although the precise mechanisms underlying these naloxone effects on volley duration and frequency are not known, they may be related to the blockade of the reported hyperpolarizing effects of opioids. Opioid peptides hyperpolarize neurons either directly by increasing membrane conductance of the postsynaptic sites (Pepper & Henderson, 1980) or indirectly by suppressing the release of neurotransmitters from presynaptic nerve terminals (Morita & North, 1981). The notion that an increase in volley

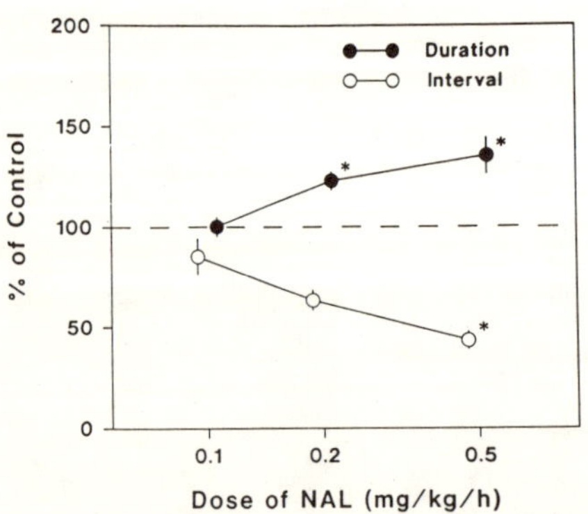

Fig. 3.   Effects of infusion of increasing doses of naloxone (NAL) on the duration and interval of MUA volleys. Each point and vertical bar represent the overall mean and SEM ($n=4$), respectively, of the percent change in each animal from its pretreatment control. *$P < 0.01$ *vs.* control. (From Kimura *et al.*, 1991).

duration is associated with membrane depolarization caused by naloxone is supported by the finding that injecting a depolarizing current prolongs the duration of spontaneous bursts of activity in magnocellular neurosecretory neurons (Andrew & Dudek, 1984). Andrew (1987) further reported that the phasic bursts in these neurons appear to be intrinsic to the neurons themselves, and contingent upon two depolarizing events: a slow depolarization between bursts which brings the membrane potential to burst threshold and thus determines the interval, and a spike depolarizing after-potential which can sum to form a plateau potential that sustains the duration of firing. Based on the analogy of magnocellular neurosecretory neurons, an increase in the excitability (*i.e.*, membrane depolarization) of neurons comprising the LHRH pulse generator could affect both the duration and frequency of MUA volleys.

These results suggest that, in the rat, endogenous opioid peptides have a tonic inhibitory effect on LHRH pulse generator activity and are profoundly involved in the control of the frequency of pulsatile LH secretion even in the absence of gonadal steroids. It is also shown for the first time that, when the inhibitory tone of opioid peptides is removed, the LHRH pulse generator can generate electrical signals in much higher frequency than those observed in ovariectomized animals.

In contrast to the rat, endogenous opioid peptides do not appear to be involved in the inhibition of the LHRH pulse generator activity in the ovariectomized monkey. Although opioid peptides have been suggested to mediate the inhibitory effects of sex steroids (Ferin *et al.*, 1984; Williams *et al.*, 1988) and corticotropin-releasing factor (CRF) (Gindoff & Ferin, 1987; Williams *et al.*, 1990a), which presumably mediates the stress effect (Petraglia *et al.*, 1986), on the pulse generator, naloxone has no effect on either LH pulses nor MUA volleys in the absence of sex steroids or CRF (Kesner *et al.*, 1986b). Opioid peptides involved in the control of the LHRH pulse generator activity seem to be continuously released in the rat brain but not in the monkey brain after ovariectomy. Differences between the rat and monkey in the effects of naloxone might be due to differences in species and/or in the time after ovariectomy, since it has been reported that the responses of LH secretion to naloxone diminish with time following ovariectomy (Massoto & Negro-Vilar, 1988; Funabashi *et al.*, 1990).

Thus, through the use of the MUA as an index, physiological significance of the opioid peptidergic system in the control of the frequency of LH pulses has been demonstrated. The opioid peptidergic system may act as a final common pathway to inhibit LHRH pulse generator activity by mediating the effects of, for example, stress, CRF and sex steroids with a resultant decrease in the frequency of pulsatile LHRH secretion.

NORADRENERGIC SYSTEM AND THE LHRH PULSE GENERATOR

Among the neurotransmitters involved in the central regulation of gonadotropin secretion, norepinephrine has been most extensively studied, and it is well established that alterations in brain norepinephrine transmission affect LH secretion (Ramirez *et al.*, 1984). Concerning the receptor specificity of norepinephrine actions, although a facilitatory effect of the activation of α-adrenergic receptors on the induction of gonadotropin surge in proestrous or sex steroid-treated ovariectomized rats has been well documented, the role of β-adrenergic receptors in this regard is still the subject of controversy. Inhibitory (Leung *et al.*, 1982), facilitatory (Al-Hamood *et al.*, 1985) and no effects (Kalra *et al.*, 1972) of the β-adrenergic system on LH surge have been reported in proestrous or sex steroid-primed ovariectomized animals.

As concerns the pulsatile LH secretion, there is general agreement that the α-adrenergic system has a facilitative role (Ramirez *et al.*, 1984), though an inhibitory effect of an intraventricularly administered α-adrenergic receptor agonist has also been reported (Leung *et al.*, 1982). The β-adrenergic system is suggested to have inhibitory (Leung *et al.*, 1982) or no effects (Weick, 1978; Bergen & Leung, 1986) on LH pulses. In the ovariectomized monkey, it has been reported that α-, but not β-, adrenergic receptor antagonists injected intravenously suppress both LH pulses and MUA volleys (Kaufman *et al.*, 1985).

The effect of both α- and β-adrenergic receptor antagonists on LHRH pulse generator activity in the ovariectomized rat was reexamined using the hypothalamic MUA as an index (Nishihara *et al.*, 1991). An α- or β-adrenergic receptor antagonist, *i.e.*, phenoxybenzamine or propranolol, was intravenously injected at a dose of 5 mg/kg 5–10 min after the end of the MUA volley. The representative effects of these drugs on the volleys and serum LH profiles are shown in Fig. 4. As summarized in Table 1, phenoxybenzamine significantly increased the interval between the MUA volleys and inhibited the appearance of the next volley for as long as 50 min. Propranolol also increased the interval between the MUA volleys, though phenoxybenzamine exhibited a slightly greater potency than propranolol. These changes in the interval of MUA volleys were faithfully reflected by the pulsatile secretion of LH. Thus, both adrenergic receptor antagonists have a potent inhibitory effect on the LHRH pulse generator activity. However, the mechanisms of the action of these antagonists might be different, since it has been observed that phenoxybenzamine often imme-

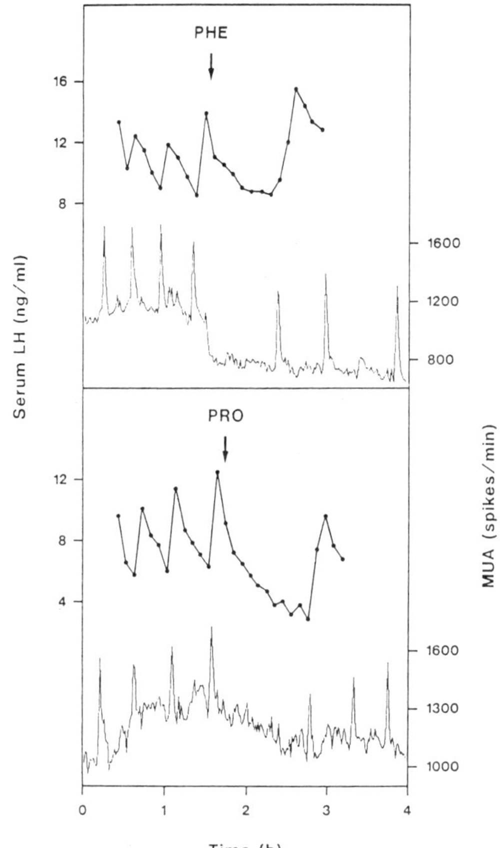

Fig. 4. Effects of a single injection of phenoxybenzamine (PHE, 5 mg/kg i.v., upper panel) and propranolol (PRO, 5 mg/kg i.v., lower panel) on LH pulses and MUA volleys in freely moving ovariectomized rats. Blood samples were taken every 6 min. (From Nishihara *et al.*, 1991).

diately reduced the baseline activity while the effect of propranolol on it was not as evident.

These results suggest that the central noradrenergic system has a tonic facilitatory effect on the electrical activity of the LHRH pulse generator through both α- and β-adrenergic receptors, and is involved in the facilitation of the frequency of LH pulses. Although there is a general agreement that the α-adrenergic system has facilitatory effect on pulsatile LH secretion, the role of β-adrenergic system is still the subject of controversy as mentioned above (Ramirez *et al.*, 1984). However, these MUA studies

TABLE 1

Effects of phenoxybenzamine (PHE), propranolol (PRO) and naloxone (NAL) on the interval between MUA volleys

Treatment	n	Pre-treatment interval		
		-3rd	-2nd	-1st
PHE	8	22.4 ± 2.1^a	20.5 ± 1.3^a	20.9 ± 1.4^a
PRO	6	21.8 ± 2.4^a	24.8 ± 2.0^a	25.8 ± 2.8^a
NAL	6	$18.0 \pm 0.5^{a,b}$	$18.8 \pm 0.8^{a,b}$	21.0 ± 1.5^a
PHE+NAL	5	22.6 ± 2.7^a	21.6 ± 2.7^a	21.2 ± 1.3^a
PRO+NAL	5	25.2 ± 3.2^a	23.4 ± 1.1^a	23.2 ± 1.2^a

Treatment	n	Post-treatment interval		
		1st*	2nd	3rd
PHE	8	59.3 ± 4.2^c	37.3 ± 4.2^b	33.3 ± 4.8^b
PRO	6	52.2 ± 7.4^b	29.3 ± 2.1^a	26.3 ± 2.1^a
NAL	6	11.2 ± 0.6^c	$17.3 \pm 1.7^{a,b}$	$13.7 \pm 3.6^{b,c}$
PHE+NAL	5	109.0 ± 25.4^b	39.0 ± 2.4^a	30.6 ± 3.6^a
PRO+NAL	5	30.6 ± 8.4^a	23.2 ± 7.6^a	18.6 ± 6.0^a

Data are means $\pm$ SEM. Values having the same superscript are not significantly different ($P > 0.05$) in each treatment group.

* The interval between MUA volleys just prior to and after drug injection.

  (From Nishihara *et al.*, 1991).

clearly demonstrated a facilitatory role of the β-adrenergic system in the control of the hypothalamic LHRH pulse generator activity. In support of this, a β-adrenergic agonist isoproterenol has been shown to affect the single unit activity of hypothalamic neurons *in vitro* (Nishihara & Kimura, 1987a). Moreover, the presence of β-adrenergic receptors in the hypothalamus has been well documented (Alexander *et al.*, 1975; Petrovic *et al.*, 1983). The inhibitory effect of propranolol on MUA volleys and LH pulses is not ascribable to systemic hypotension, since the blockade of opioid peptidergic transmission by naloxone can induce both MUA volleys and LH pulses even in the presence of propranolol as described below. Although epinephrine also acts through β-adrenergic receptors and plays a prominent role in inducing LH surge in ovariectomized rats treated with ovarian steroids (Crowley & Terry, 1981), the inhibition of epinephrine sysnthesis has been shown to have no effect on pulsatile secretion of LH (Crowley *et al.*, 1982). Thus, norepinephrine itself might facilitate the excitability of the LHRH pulse generator through both α- and β-adrenergic receptors.

The reason for the discrepancy between the results described here and those in other studies (Weick, 1978; Bergen & Leung, 1986), which failed to show an effect of propranolol on LH pulses, is unclear since comparable doses (5–10 mg/kg) of propranolol were intravenously administered in adult ovariectomized rats in all these studies. Differences in the timing of drug injection and/or methods for the pulse determination could account for the difference. By monitoring the MUA, the drugs can be injected at precise and constant intervals after the onset of the MUA volley, and the exact interval between the volleys can be estimated.

The MUA study has demonstrated that the blockade of norepinephrine transmissions *via* either α- or β-adrenergic receptors reduces the frequency of periodic activation of the LHRH pulse generator. This seems to be incompatible with the notion of Terasawa *et al.* (1988) that norepinephrine release from the stalk-median eminence of ovariectomized monkeys is pulsatile and is synchronous with pulsatile LHRH release, since the pharmacological blockade of postsynaptic receptors would not be expected to change the frequency of periodic noradrenergic neuronal activity. It may yet be possible that, when periodic noradrenergic signals are available, the activity of the LHRH pulse generator depends on them, but when such signals are not available, the pulse generator fires intrinsically or in association with other neuronal signals. However, it has been shown in ovariectomized rats that pulsatile LH release, abolished after acute norepinephrine depletion, can be reestablished for several hours by a single injection of an α-adrenergic agonist, clonidine (Estes *et al.*, 1982). Changes in the noradrenergic tone appear to play a significant role in the modulation of the frequency of LHRH pulses. Functional differences between the α- and β-adrenergic systems in controlling the LHRH pulse generator activity remain to be resolved.

INTERACTION BETWEEN OPIOID PEPTIDERGIC AND NORADRENERGIC SYSTEMS IN CONTROLLING THE LHRH PULSE GENERATOR

As described above, the electrical activity of the LHRH pulse generator in ovariectomized rats is under the influence of inhibitory opioid peptidergic and facilitatory noradrenergic tones. Do these two neuronal systems act independently or interact in controlling the electrical activity of the LHRH pulse generator? It has been proposed that some effects of opioid peptides on LH surge in castrated animals treated with ovarian steroids and on basal LH secretion in intact animals of both sexes are mediated by norepinephrine (Kalra & Simpkins, 1981; Van Vugt *et al.*, 1981; Blank & Bohnet, 1983). In these reports, opioid peptides are suggested to inhibit LH secretion by suppressing norepinephrine release in the brain. Other studies, however, have suggested that opioid peptides did not require noradrenergic mediation (Miller *et al.*, 1985) or that the opioid receptor antagonist and the adrenergic receptor agonist acted synergistically (Clough *et al.*, 1990). With regard to the control of pulsatile LH secretion, it remains unknown how the inhibitory opioid peptidergic effect and facilitatory noradrenergic effect interact. Thus, a study was performed to elucidate whether norepinephrine mediated the opioid effect on LHRH pulse generator activity (Nishihara *et al.*, 1991).

In this experiment, a bolus injection of a much larger dose of naloxone (2 mg/kg) than that in previous infusion experiment mentioned above was applied to see if naloxone could produce frequent MUA volleys even after the blockade of noradrenergic transmissions. Single injection of naloxone decreased the intervals between the MUA volleys (Fig. 5), though the

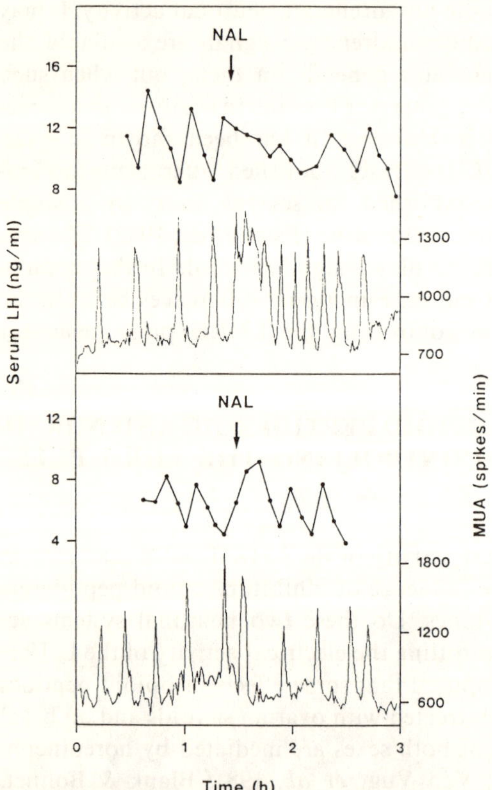

Fig. 5.   Effects of a single injection of naloxone (NAL, 2 mg/kg i.v.) on LH pulses and MUA volleys in freely moving ovariectomized rats. Blood samples were taken every 6 min. (From Nishihara *et al.*, 1991).

responses to naloxone markedly varied from animal to animal. In some animals, naloxone produced an MUA volley of rather long duration followed by volleys at short intervals (Fig. 5, upper panel), while in others it induced only one immediate MUA volley (Fig. 5, lower panel). What was common to all the animals studied was that an MUA volley appeared immediately after a naloxone injection (within 2 min), and thus the interval between the volleys just prior to and after the injection was significantly decreased (Table 1).

The effects of naloxone injected 10 min after the injection of phenoxybenzamine and propranolol on MUA volleys and serum LH profiles are

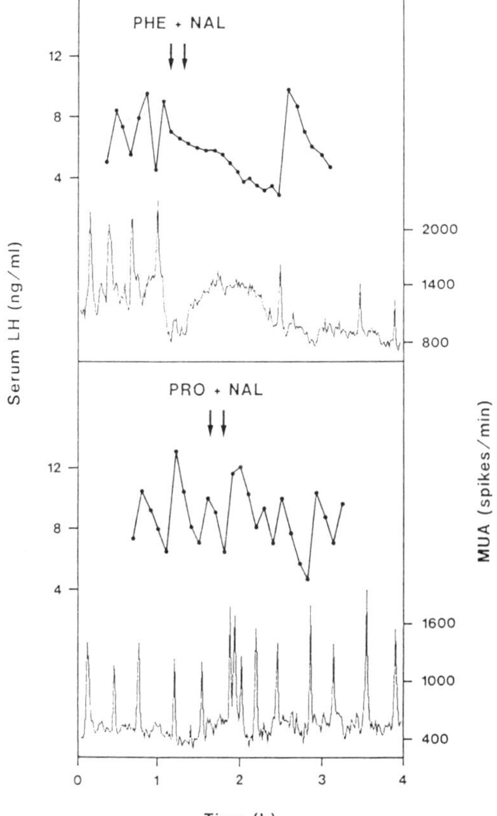

Fig. 6. Effects of naloxone (NAL, 2 mg/kg i.v.) injected 10 min after the injection of phenoxybenzamine (PHE, 5 mg/kg i.v., upper panel) or propranolol (PRO, 5 mg/kg i.v., lower panel) on LH pulses and MUA volleys in freely moving ovariectomized rats. Blood samples were taken every 6 min. (From Nishihara *et al.*, 1991).

shown in Fig. 6, and those on the intervals between MUA volleys are summarized in Table 1. Naloxone given 10 min after the injection of propranolol induced the volleys associated with LH pulses with a latency of a few minutes. However, when given 10 min after the injection of phenoxybenzamine, naloxone failed to induce such immediate MUA volleys. This suggests that the action of endogenous opioid peptides on the LHRH pulse generator activity, and, in turn, on the frequency of LH pulses, requires an α-, but not β-adrenergic receptor mediated mechanism. Thus, the increase in the frequency of MUA volleys caused by naloxone is probably due to the increase in α-adrenergic receptor mediated synaptic transmissions resulting from the withdrawal of opioid peptidergic suppression. Contrarily, the β-adrenergic system appears to act rather independently on the LHRH pulse generator. The observation that MUA volleys and LH pulses can be induced even under β-, but not α-adrenergic receptor blockade also suggests that the α-adrenergic system may be a predominant adrenergic system involved in the control of pulse generator activity.

It has been reported that naloxone stimulates the release of norepinephrine from the nerve terminals in the hypothalamic fragments *in vitro* (Leadem *et al.*, 1985; Rasmussen *et al.*, 1988) or the turnover rate of norepinephrine in the hypothalamic area *in vivo* (Adler & Crowley, 1984). Naloxone implanted in the medial preoptic area or the median eminence-arcuate region has also been shown to stimulate LH release in ovariectomized steroid-treated rats (Kalra, 1981). Further, a principal role of the opioid peptidergic system is regarded as presynaptic inhibition in the central nervous system (Motita & North, 1981). These suggest that, although opioid receptors are located both on the nerve terminals and cell bodies of noradrenergic neurons (Pert *et al.*, 1976; Langer, 1981), endogenous opioid peptides may act mainly on the nerve terminals in the hypothalamus to control the release of norepinephrine, which, in turn, controls LHRH pulse generator activity.

The concept of a noradrenergic mediation of the opioid effect on LH secretion is thus further supported by the MUA study. However, the possibility of a direct opioid effect on the LHRH pulse generator cannot be definitively excluded. Direct synaptic contacts have been demonstrated between opioid- and LHRH-containing neurons in the medial basal hypothalamus of the monkey (Thind & Goldsmith, 1988). Nevertheless, the MUA study suggests the noradrenergic receptor specificity in the interaction between the noradrenergic and opioid peptidergic systems in the control of LHRH pulse generator activity.

CONCLUSION

Hypothalamic MUA recordings provide direct access to the central component of the neuroendocrine control system which governs reproductive function. The MUA studies have shown that the opioid peptidergic and noradrenergic systems have tonic inhibitory and facilitatory effects, respectively, on the electrical activity of the LHRH pulse generator. In other words, the former system decreases and the latter system increases the frequency of LH pulses, which is one of the key determinants of the gonadal function. Thus, frequency modulation is a principal regulatory mode underlying the central control of gonadotropin secretion. Digital signals, as well as analog signals, may have important physiological significance in the neuroendocrine control of reproductive function.

The effects of many environmental factors on reproductive function are mediated by opioid peptidergic and noradrenergic systems. Neither of these neuronal systems itself, however, appears to be the LHRH pulse generator but both appear to be modulators of this generator. In support of this, recovery of LH pulses after complete deafferentation of the hypothalamus (Blake & Sawyer, 1974; Ohkura *et al.*, 1991) or disruption of the ascending noradrenergic fibers (Soper & Weick, 1980) has been reported. It has also been shown that the response of LH secretory pattern to naloxone changes with time after ovariectomy (Massoto & Negro-Vilar, 1988; Funabashi *et al.*, 1990). These studies suggest that both the opioid peptidergic and noradrenergic systems play a permissive role in the control of pulsatile LH secretion.

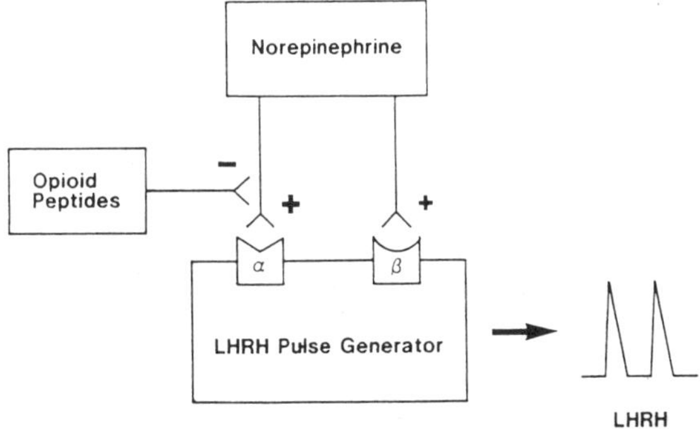

Fig. 7.   Schematic diagram summarizing the interactions between the noradrenergic and opioid peptidergic systems in controlling the LHRH pulse generator activity.

One of the main questions to be resolved is the relationship between the generating systems of LH pulse and preovulatory LH surge, which is also under the influence of endogenous opioid peptides and norepinephrine. Although information regarding electrical events operating LH surge is limited at present, a marked decline in the LHRH pulse generator activity during LH surge has been observed in the monkey (Kesner *et al.*, 1987; O'Byrne *et al.*, 1990), suggesting a major distinction between the mechanisms generating LH pulse and LH surge in this species. The MUA study on other species will promote a better understanding of the organization of the LH surge generating system and its species differences.

In conclusion, as schematically illustrated in Fig. 7, the MUA studies described here suggest that norepinephrine increases the frequency of the electrical activity of the LHRH pulse generator *via* both α- and β-adrenergic receptors, and that opioid peptides decrease its frequency through the inhibition of norepinephrine release from the nerve terminals, the action of which is mediated *via* α-adrenergic receptors. Although it has been reported that many other neuronal systems are also involved in the control of the pulse generator activity, what kind of neurons and mechanisms are involved in the phasic bursts of LHRH neurons is not known. Further studies are needed to clarify the cellular mechanism of the LHRH pulse generating system.

ACKNOWLEDGEMENT

The authors are deeply indebted to Dr. E. Knobil, the University of Texas, for his major contribution to development of the MUA recording systems for the rat and goat.

REFERENCES

Adler, B.A. & Crowley, W.R. (1984). Modulation of luteinizing hormone releasing hormone release and catecholamine activity by opiates in the female rat. *Neuroendocrinology* **38**, 248–253.

Alexander, R.W., Davis, J.N. & Leekowitz, R.J. (1975). Direct identification and characterization of β-adrenergic receptors in rat brain. *Nature* **258**, 437–439.

Al-Hamood, M.H., Gilmore, D.P. & Wilson, C.A. (1985). Evidence for stimulatory β-adrenergic component in the release of the ovulatory LH surge in pro-oestrous rats. *Journal of Endocrinology* **106**, 143–151.

Allen, L.G. & Kalra, S.P. (1986). Evidence that a decrease in opioid tone may evoke preovulatory luteinizing hormone release in the rat. *Endocrinology* **118**, 2375–2381.

Andrew, R.D. (1987). Endogenous bursting by rat supraoptic neuroendocrine cells is calcium dependent. *Journal of Physiology* **384**, 451–465.

Andrew, R.D. & Dudek, F.E. (1984). Analysis of intracellularly recorded phasic bursting by mammalian neuroendocrine cells. *Journal of Neurophysiology* **51**, 552–566.

Bergen, H. & Leung, P.C.K. (1986). Norepinephrine inhibition of pulsatile LH release: Receptor specificity. *American Journal of Physiology* **250**, E205–E211.

Bhanot, R. & Wilkinson, M. (1983). Opiatergic control of LH secretion is eliminated by gonadectomy. *Endocrinology* **112**, 399–401.

Blake, C.A. & Sawyer, C.H. (1974). Effects of hypothalamic deafferentation on the pulsatile rhythm in plasma concentrations of luteinizing hormone in ovariectomized rats. *Endocrinology* **94**, 730–736.

Blank, M.S. & Bohnet, H.G. (1983). α-adrenergic but β-adrenergic or serotonergic antagonists block naloxone-induced LH release. *Neuroendocrinological Letters* **5**, 3–8.

Clarke, I.J. & Cummins, J.T. (1982). The temporal relationship between gonadotropin releasing hormone (GnRH) and luteinizing hormone (LH) secretion in ovariectomized ewes. *Endocrinology* **111**, 1737–1739.

Clarke, I.J., Cummins, J.T., Findlay, J.K., Burman, K.J. & Doughton, B.W. (1984). Effects on plasma luteinizing hormone and follicle-stimulating hormone of varying the frequency and amplitude of gonadotropin-releasing hormone pulses in ovariectomized ewes with hypothalamo-pituitary disconnection. *Neuroendocrinology* **39**, 214–221.

Clough, R.W., Hoffman, G.E. & Sladek, C.D. (1990). Synergistic interaction between opioid receptor blockade and α-adrenergic stimulation on luteinizing hormone-releasing hormone (LHRH) secretion *in vitro*. *Neuroendocrinology* **51**, 131–138.

Crowley, W.R. & Terry, L.C. (1981). Effects of an epinephrine synthesis inhibitor, SKF64139, on the secretion of luteinizing hormone in ovariectomized female rat. *Brain Research* **204**, 231–235.

Crowley, W.R., Terry, L.C. & Johnson, M.D. (1982). Evidence for the involvement of central epinephrine system in the regulation of luteinizing hormone, prolactin, and growth hormone release in female rat. *Endocrinology* **110**, 1102–1107.

Devorshak-Harvey, E., Bona-Gallo, A. & Gallo, R.V. (1987). Endogenous opioid peptide regulation of pulsatile luteinizing hormone secretion during pregnancy in the rat. *Neuroendocrinology* **46**, 369–378.

Dierschke, D.J., Bhattacharya, A.N., Atkinson, L.E. & Knobil, E. (1970). Circhoral oscillations of plasma LH levels in the ovariectomized rhesus monkey. *Endocrinology* **87**, 850–853.

Estes, K.S., Simpkins, J.W. & Kalra, S.P. (1982). Resumption with clonidine of pulsatile LH release following acute norepinephrine depletion in ovariectomized rats. *Neuroendocrinology* **35**, 56–62.

Ferin, M., Van Vugt, D. & Wardlaw, S. (1984). The hypothalamic control of the menstrual cycle and the role of endogenous opioid peptides. *Recent Progress in Hormone Research* **40**, 441–485.

Ferin, M., Wehrenberg, W.B., Lam, N.Y., Alston, E.J. & Vande Wiele, R.L. (1982). Effects and site of action of morphine on gonadotropin secretion in the female rhesus monkey. *Endocrinology* **111**, 1652–1656.

Funabashi, T., Kato, A. & Kimura, F. (1990). Naloxone affects LH secretory pattern in the short and long-term ovariectomized rat. *Neuroendocrinology* **52**, 35–41.

Garcia-Segura, L.M., Baetens, D. & Naftolin, F. (1986). Synaptic remodelling in arcuate nucleus after injection of estradiol valerate in adult female rats. *Brain Research* **366**, 131–136.

Gay, V.L. & Plant, T.M. (1987). N-methyl-D,L-aspartate elicits hypothalamic gonadotropin releasing hormone release in prepubertal male rhesus monkeys (*Macaca mulatta*). *Endocrinology* **120**, 2289–2296.

Gindoff, P.R. & Ferin, M. (1987). Endogenous opioid peptides modulate the effects of corticotropin releasing factor on gonadotropin release in the primate. *Endocrinology* **121**, 837–842.

Hiruma, H., Nishihara, M. & Kimura. F. (1992). Hypothalamic electrical activity that relates to the pulsatile release of luteinizing hormone exhibits diurnal variation in ovariectomized rats. *Journal of Neuroendocrinology*, submitted.

Kalra, S.P. (1981). Neural loci involved in naloxone-induced luteinizing hormone release: Effects of a norepinephrine synthesis inhibitor. *Endocrinology* **109**, 1805–1810.

Kalra, P.S., Kalra, S.P., Krulich, L., Fawcett, C.P. & McCann, S.M. (1972). Involvement of norepinephrine in transmission of the stimulatory influence of progesterone on gonadotropin release. *Endocrinology* **90**, 1168–1176.

Kalra, S.P. & Simpkins, J.W. (1981). Evidence for noradrenergic mediation of opioid effects on luteinizing hormone secretion. *Endocrinology* **109**, 776–782.

Kamel, F., Balz, J.A., Kubajak, C.L. & Schneider, V.A. (1987). Effects of luteinizing hormone (LH)-releasing hormone pulse amplitude and frequency on LH secretion by perfused rat anterior pituitary cells. *Endocrinology* **120**, 1644–1650.

Kaufman, J.-M., Kesner, J.S., Wilson, R.C. & Knobil, E. (1985). Electrophysiological manifestation of luteinizing hormone-releasing hormone pulse generator activity in the rhesus monkey: Influence of α-adrenergic and dopaminergic blocking agents. *Endocrinology* **116**, 1327–1333.

Kawakami, M., Uemura, T. & Hayashi, R. (1982). Electrophysiological correlates of pulsatile gonadotropin release in rats. *Neuroendocrinology* **35**, 63–67.

Kawano, H. & Daikoku, S. (1981). Immunohistochemical demonstration of LHRH neurons and their pathways in the rat hypothalamus. *Neuroendocrinology* **32**, 179–186.

Kaynard, A.H., Pau, K.-Y. F., Hess, D.L. & Spies, H.G. (1990). Third-ventricular infusion of neuropeptide Y suppresses luteinizing hormone secretion in ovariectomized rhesus monkey. *Endocrinology* **127**, 2437–2444.

Kesner, J.S., Kaufman, J.-M., Wilson, R.C., Kuroda, G. & Knobil, E. (1986a). On the short-loop feedback regulation of the hypothalamic luteinizing hormone releasing hormone 'pulse generator' in the rhesus monkey. *Neuroendocrinology* **42**, 109–111.

Kesner, J.S., Kaufman, J.-M., Wilson, R.C., Kuroda, G. & Knobil, E. (1986b). The effect of morphine on the electrical activity of the hypothalamic luteinizing hormone-releasing hormone pulse generator in the rhesus monkey. *Neuroendocrinology* **43**, 686–688.

Kesner, J.S., Wilson, R.C., Kaufman, J.-M., Hotchkiss, J., Chen, Y., Yamamoto, H., Pard, R.R. & Knobil, E. (1987). Unexpected responses of the hypothalamic gonadotropin-releasing hormone "pulse generator" to physiological estradiol inputs in the absence of the ovary. *Proceedings of the National Academy of Sciences, U.S.A.* **84**, 8745–8749.

Kimura, F., Nishihara, M., Hiruma, H. & Funabashi, T. (1991). Naloxone increases the frequency of the electrical activity of luteinizing hormone-releasing hormone pulse generator in long-term ovariectomized rats. *Neuroendocrinology* **53**, 97–102.

Kinoshita, F., Nakai, Y., Katakami, H., Kato, Y., Yajima, H. & Imura, H. (1980). Effect of β-endorphin on pulsatile LH release in conscious castrated rats. *Life Sciences* **27**, 843–846.

Knobil, E. (1980). The neuroendocrine control of the menstrual cycle. *Recent Progress in Hormone Reserch* **36**, 53–88.

Kubo, K., Kiyota, Y. & Fukunaga, S. (1983). Effects of third ventricular injection of β-endorphine on luteinizing hormone surges in female rat: Sites and mechanisms of opioid actions in the brain. *Endocrinologia Japonica* **30**, 419–433.

Lamberts, R., Vijayan, E., Graf, M., Mansky, T. & Wuttke, W. (1983). Involvement of preoptic-anterior hypothalamic GABA neurons in the regulation of pituitary LH and prolactin release. *Experimental Brain Research* **52**, 356–362.

Langer, S.Z. (1981). Presynaptic regulation of the release of catecholamines. *Pharmacological Review* **32**, 337–362.

Leadem, C.A., Crowley, W.R., Simpkins, J.W. & Kalra, S.P. (1985). Effects of naloxone on catecholamine and LHRH release from the perifused hypothalamus of the steroid-primed rat. *Neuroendocrinology* **40**, 497–500.

Leung, P.C.K., Arendash, G.W., Whitmoyer, D.I., Gorski, R.A. & Sawyer, C.H. (1982). Differential effects of central adrenoceptor agonists on luteinizing hormone release. *Neuroendocrinology* **34**, 207–214.

Lincoln, D.W. (1988). LHRH: Pulse generation. In: *Pulsatility in Neuroendocrine Systems* (Leng, G., ed.), CRC Press, Florida, pp. 35–60.

Lincoln, G.A. & Short, R.V. (1980). Seasonal breeding: Nature's contraceptive. *Recent Progress in Hormone Research* **36**, 1–52.

Massoto, C. & Negro-Vilar, A. (1988). Gonadectomy influences the inhibitory effect of the endogenous opiate system on pulsatile gonadotropin secretion. *Endocrinology* **123**, 747–752.

Miller, M.A., Clifton, D.K. & Steiner, R.A. (1985). Noradrenergic and endogenous opioid pathways in the regulation of luteinizing hormone secretion in the male rat. *Endocrinology* **117**, 544–548.

Mori, Y., Nishihara, M., Tanaka, T., Shimizu, T., Yamaguchi, M., Takeuchi, Y. & Hoshino, K. (1991). Chronic recording of electrophysiological manifestation of the hypothalamic gonadotropin-releasing hormone pulse generator activity in the goat. *Neuroendocrinology* **53**, 392–395.

Morita, K. & North, R.A. (1981). Opiates and enkephalin reduce the excitability of neural processes. *Neuroscience* **6**, 1943–1951.

Nishihara, M., Hiruma, H. & Kimura, F. (1991). Interactions between the noradrenergic and opioid peptidergic systems in controlling the electrical activity of luteinizing hormone-releasing hormone pulse generator in ovariectomized rats. *Neuroendocrinology* **54**, 321–326.

Nishihara, M. & Kimura. F. (1987a). Differential influences of adrenoceptor agonists on the excitability of arcuate and medial preoptic neurons. *Japanese Journal of Veterinary Science* **49**, 1154–1156.

Nishihara, M. & Kimura, F. (1987b). Roles of γ-aminobutyric acid and serotonin in the arcuate nucleus in the control of prolactin and luteinizing hormone secretion. *Japanese Journal of Physiology* **37**, 955–961.

Nishihara, M. & Kimura, F. (1989). Postsynaptic effects of prolactin and estrogen on arcuate neurons in rat hypothalamic slices. *Neuroendocrinology* **49**, 215–218.

Nishihara, M., Matsukawa, T. & Kimura, F. (1986). Responses of arcuate neurons to some putative neurotransmitters in perfused rat hypothalamic slices: Effects of *in vivo* and *in vitro* estrogen treatments. *Japanese Journal of Physiology* **36**, 683–697.

O'Byrne, K., Thalabard, J.-C., Williams, C.L., Grosser, P.M., Hotchkiss, J., Nishihara, M., Pate, S. & Knobil, E. (1990). An unexpected decline in GnRH pulse generator frequency in the late follicular phase of the rhesus monkey menstrual cycle. *Abstracts of 72nd Annual Meeting of the Endocrine Society*, Atlanta, p. 289.

Ohkura, S., Tsukamura, H. & Maeda, K.-I. (1991). Effects of various types of hypothalamic deafferentation on luteinizing hormone pulses in ovariectomized rats. *Journal of Neuroendocrinology* **3**, 503–508.

Pang, C.N., Zimmermann, E. & Sawyer, C.H. (1977). Morphine inhibition of the preovulatory surges of plasma luteinizing hormone and follicle stimulating hormone in the rat. *Endocrinology* **101**, 1726–1732.

Pepper, C.M. & Henderson, G.H. (1980). Opiates and opioid peptides hyperpolarize locus coeruleus neurons *in vitro*. *Science* **209**, 394–396.

Pert, C.B., Kuhar, M.J. & Snyder, S.H. (1976). Opiate receptor: Autoradiographic localization in rat brain. *Proceedings of the National Academy of Sciences, U.S.A.* **73**, 3729–3733.

Petraglia, F., Locatelli, V., Penalva, A., Cocchi, D., Genazzani, A.R. & Muller, E.E. (1984). Gonadal steroid modulation of naloxone-induced LH secretion in the rat. *Journal of Endocrinology* **101**, 33–39.

Petraglia,F., Vale, W. & Rivier, C. (1986). Opioids act centrally to modulate stress-induced decrease in luteinizing hormone in the rat. *Endocrinology* **119**, 2445–2450.

Petrovic, S.L., McDonald, J.K., Snyder, G.D. & McCann, S.M. (1983). Characterization of β-adrenergic receptors in rat brain and pituitary using a new high-affinity [^{125}I]-iodocyanopindolol. *Brain Research* **261**, 249–259.

Piva, F., Maggi, P., Limonta, p., Motta, M. & Martini, L. (1985). Effect of naloxone on luteinizing hormone, follicle-stimulating hormone, and prolactin secretion in the different phases of the estrous cycle. *Endocrinology* **117**, 766–772.

Pohl, C.R., Richardson, D.W., Hutchinson, J.S., Germak, J.A. & Knobil, E. (1983). Hypophysiotropic signal frequency and the functioning of the pituitary-ovarian system in the rhesus monkey. *Endocrinology* **112**, 2076–2080.

Ramirez, V.D., Feder, H.H. & Sawyer, C.H. (1984). The role of brain catecholamines in the regulation of LH secretion: A critical inquiry. In: *Frontiers in Neuroendocrinology*, Vol. 8 (Martini, L. and Gannong, W.F., eds.), Raven Press, New York, pp. 27-84.

Rasmussen, D.D., Kennedy, B.P., Ziegler, M.G. & Nett, T.M. (1988). Endogenous opioid inhibition and facilitation of gonadotropin releasing hormone release from the median eminence *in vitro*: Potential role of catecholamines. *Endocrinology* **123**, 2916–2921.

Schulz, R., Willhelm, A., Pirke, K.M., Gramsch, C. & Hers, A. (1981). β-endorphin and dynorphin control serum luteinizing hormone level in immature female rats. *Nature* **294**, 757–759.

Soper, B.D. & Weick, R.F. (1980). Hypothalamic and extrahypothalamic mediation of pulsatile discharges of luteinizing hormone in the ovariectomized rat. *Endocrinology* **106**, 348–355.

Terasawa, E., Krook, C., Hei, D.L., Gearing, M., Schultz, N.J. & Davis, G.A. (1988). Norepinephrine is a possible neurotransmitter stimulating pulsatile release of luteinizing hormone-releasing hormone in the rhesus monkey. *Endocrinology* **123**, 1808–1816.

Thind, K.K. & Goldsmith, P.C. (1988). Infundibular gonadotropin-releasing hormone neurons are inhibited by direct opioid and autoregulatory synapses in juvenile monkeys. *Neuroendocrinology* **47**, 203–216.

Van Vugt, D.A., Aylsworth, C.F., Sylvester, P.W., Leung, F.C. & Meites, J. (1981).

Evidence for hypothalamic noradrenergic involvement in naloxone-induced stimulation of luteinizing hormone release. *Neuroendocrinology* **33**, 261–264.

Weick, R.F. (1978). Acute effects of adrenergic receptor blocking drugs and neuroleptic agents on pulsatile discharges of luteinizing hormone in the ovariectomized rat. *Neuroendocrinology* **26**, 108–117.

Wildt, L., Hausler, A., Marshall, G., Hutchison, J.S., Plant, T.M., Belchetz, P.E. & Knobil, E. (1981). Frequency and amplitude of gonadotropin-releasing hormone stimulation and gonadotropin secretion in the rhesus monkey. *Endocrinology* **109**, 376–385.

Williams, C.L., Nishihara, M., Thalabard, J.-C., Grosser, P.M., Hotchkiss, J. & Knobil, E. (1988). The hypothalamic GnRH pulse generator of the rhesus monkey: The duration of phasic electrical activity and its control. *Abstracts of 18th Annual Meeting of Society for Neuroscience*, Toronto, p. 1068.

Williams, C.L., Nishihara, M., Thalabard, J.-C., Grosser, P.M., Hotchkiss, J. & Knobil, E. (1990a). Corticotropin releasing factor and gonadotropin releasing hormone pulse generator activity in the rhesus monkey: Electrophysiological studies. *Neuroendocrinology* **52**, 133–137.

Williams, C.L., Nishihara, M., Thalabard, J.-C., Grosser, P.M., Hotchkiss, J. & Knobil, E. (1990b). Duration and frequency of multiunit electrical activity associated with the hypothalamic gonadotropin releasing hormone pulse generator in the rhesus monkey: Differential effects of morphine. *Neuroendocrinology* **52**, 225–228.

Williams, C.L., Thalabard, J.-C., O'Byrne, K.T., Grosser, P.M., Nishihara, M., Hotchkiss, J. & Knobil, E. (1990c). Duration of phasic electrical activity of the hypothalamic gonadotropin-releasing hormone pulse generator and dynamics of luteinizing hormone pulses in the rhesus monkey. *Proceedings of the National Academy of Sciences, U.S.A.* **87**, 8580–8582.

Wilson, R.C., Kesner, J.S., Kaufman, J.-M., Uemura, T., Akema, T. & Knobil, E. (1984). Central electrophysiological correlates of pulsatile luteinizing hormone secretion in the rhesus monkey. *Neuroendocrinology* **39**, 256–260.

Wray, S. & Hoffman, G. (1986). A developmental study of the quantitative distribution of LHRH neurons within the central nervous system of postnatal male and female rats. *Journal of Comparative Neurology* **252**, 522–531.

5 Central Integration of Photoperiodicity for Gonadotropin Release in Ruminants

Yuji MORI

Laboratory of Veterinary Ethology, Department of Veterinary Medical Science, The University of Tokyo, Tokyo 113

Seasonal reproduction is a 'reproductive strategy' commonly seen among a wide range of mammalian species inhabiting areas extending from the temperate zone to the higher latitude where the environment shows a dramatic seasonal change. In most of these animals parturition occurs in spring when the environmental conditions become optimal for their offsprings to survive. Sheep and goats are among these species, and they are called short-day breeders because they mate and conceive in autumn. The horse and many rodents including hamsters are categorized as long-day breeders because their reproductive activity is seen mainly in spring. Despite horses and goats breeding in opposite seasons they both deliver foals and kids in spring (Fig. 1). It is therefore believed that each species has set the breeding season according to its own gestation length so that a spring birth is ensured.

This seasonal reproduction is a mechanism of "Nature's contraceptive" (Lincoln & Short, 1980), by which the animal falls into physiological hypogonadism in the anestrous season to avoid an undesirable winter birth. Among various environmental factors (*e.g.*, temperature, precipitation and vegetation) the photoperiod, namely an annual change in daylength, is the principal external cue for the institution of annual reproductive cyclicity in most mammalian species investigated.

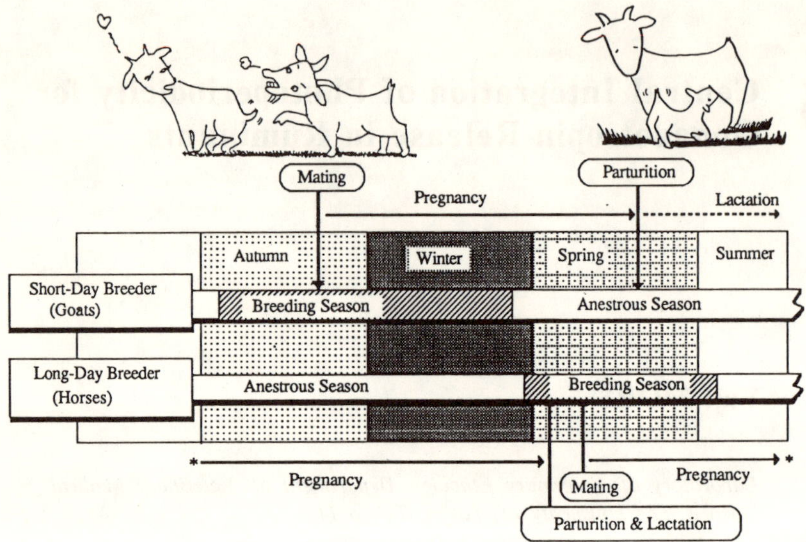

Fig. 1.   Representative annual reproductive patterns in the short-day breeder (goat) and the long-day breeder (horse). In the goat ovulatory cycles normally recur every 3 weeks in the breeding season (autumn and winter), whereas gonadal activity is arrested in the non-breeding season (spring and summer). Since the gestation period of the goat is about 5 months, they deliver kids in spring, whereas the horse foals in spring, due to the longer length of gestation (11 months). A lactating mare soon starts estrous cycles and conceives within the same spring. The female accepts a male's approach only at estrus. The endocrine sequence timing the estrus and ovulation to maximize conception efficiency is described in Fig. 2.

This article discusses the neuroendocrine mechanism underlying the seasonal reproduction by overviewing our own studies in the ruminant: (1) secretory pattern of gonadotropins in the ovulatory cycle, (2) photoperiodic control of gonadotropin secretion, and (3) role of the pineal gland in the reproductive photoperiodicity. Further detail may be found in the following reviews and articles: Lincoln & Short, 1980; Karsch *et al.*, 1984; Foster *et al.*, 1986; Maeda *et al.*, 1988; Bronson, 1988; Thiery & Martin, 1991; Mori & Maeda, 1991.

GONADOTROPIN SECRETION IN THE OVULATORY CYCLE

Secretory profiles of pituitary gonadotropins show a typical change in the preovulatory period. Here described is the endocrine sequence underlying the ovarian cyclicity in the female goat.

Ovulatory Cycles in the Female Goat

Mature female goats come into estrus every 3 weeks. The estrus lasts for a day or two when the female allows the male to mate, and the ovulation usually takes place several hours after the end of estrus (Mori & Kano, 1984). Unless the female becomes pregnant the estrous cycle appears throughout the breeding season that usually extends from early autumn to late winter, whereas the cyclicity ceases in the anestrous season, *i.e.* spring and summer (see Fig. 1). This annual pattern of reproductive activity is under photoperiodic control as will be discussed later.

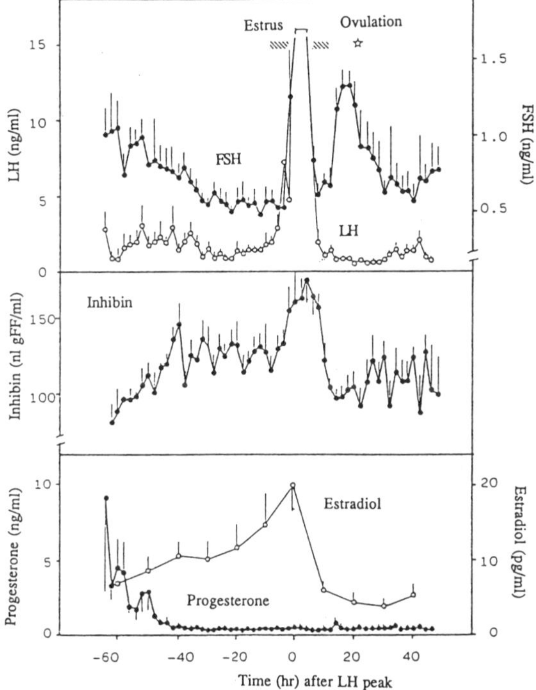

Fig. 2. Changes in plasma concentrations of gonadotropins (LH and FSH), inhibin, and ovarian steroids (progesterone and estradiol) during the follicular phase in relation to the occurrence of behavioral estrus and ovulation in female goats. Decline in circulating progesterone accompanying by the luteolysis is followed by the follicular development and a gradual increase of estradiol secretion that subsequently induces preovulatory LH/FSH surges and estrus. Inhibin levels also rise as follicles develop. Note a secondary rise in FSH but not LH preceding the ovulation. Each point depicts mean value (±SEM) for 6 goats. (From Mori & Kano, 1984; Mori *et al.*, unpublished data)

Changes in circulating levels of hormones secreted by the anterior pituitary gland and the ovary during the follicular phase of the estrous cycle in the goat are illustrated in Fig. 2. This endocrine sequence determines the time of the estrus and ovulation with an appropriate interval and mating occurs with maximum conception efficiency. Decline in circulating progesterone at luteolysis is followed by follicular development and associated with an increase of estradiol in the systemic circulation that subsequently induces preovulatory gonadotropin surges and estrous behavior (Mori & Kano, 1984; Mori *et al.*, 1987).

Pulsatile Secretion of LH and the Gonadotropin-releasing Hormone (GnRH) Pulse Generator

Like many other species investigated so far the basal patterns of LH secretion in the goat are not constant but are pulsatile in nature as shown in Fig. 3. Frequency of the LH pulse increases following a transition from

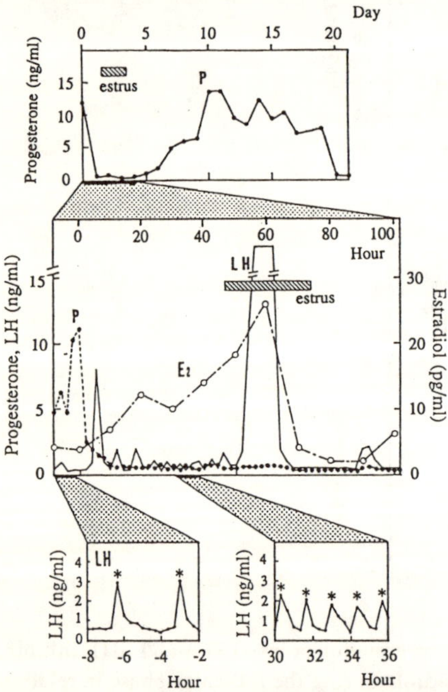

Fig. 3.   Profiles of LH secretion in a normally cycling goat. Basal patterns of LH secretion are not constant but are pulsatile as shown in the bottom two plates. The luteolysis was induced by an injection of prostaglandin $F_{2\alpha}$ at 0 hr. The frequency of LH pulse increases following a transition from the luteal to the follicular phase, and this is considered to be essential for the follicular development and subsequent preovulatory endocrine change leading to estrus and ovulation (From Tsuda *et al.*, unpublished data).

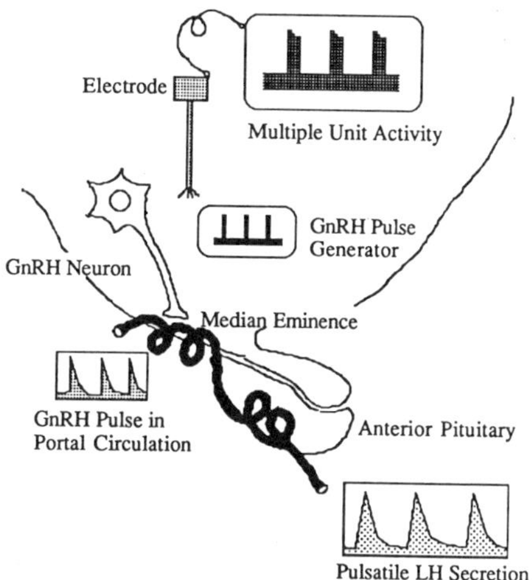

Fig. 4. Pulsatile pattern of circulating LH is caused by an intermittent discharge of hypothalamic decapeptide GnRH at the median eminence of the hypothalamus. The GnRH released into the pituitary portal circulation reaches the anterior pituitary lobe and stimulates LH secretion from gonadotrophs. The GnRH pulse generator is postulated to be a neural mechanism commanding the periodic activation of GnRH neurons and resulting neurosecretion of decapeptide from the nerve terminals. Recent advance in the electrophysiological technique has enabled a long-term recording of multiunit activity (MUA) that is specifically associated with pulsatile patterns of LH secretion from the hypothalamus of conscious and unrestrained goats as shown in Figs. 5 and 6.

the luteal phase to the follicular phase, and this increase is essential for either normal follicular development or subsequent preovulatory endocrine events leading to the onset of estrus and ovulation (Karsch *et al.*, 1984; Martin, 1984).

As illustrated in Fig. 4 the hypothalamic pulse generator which regulates intermittent discharge of GnRH into the pituitary circulation and thereby modulates the pulsatile secretion of LH has been recognized as a key determinant of reproductive function in all mammals examined to date (Knobil, 1980; Lincoln *et al.*, 1985). Although techniques were developed to monitor directly GnRH secretion in unanesthetized animals (Levin *et al.*, 1982; Clarke & Cummins, 1982), in the light of technical difficulties, most researchers are still using LH secretion as an indirect index of GnRH release, and this approach has inevitable limitations that preclude firm conclusions.

Electrophysiological Approach to the GnRH Pulse Generator

The electrophysiological manifestation of the hypothalamic GnRH pulse

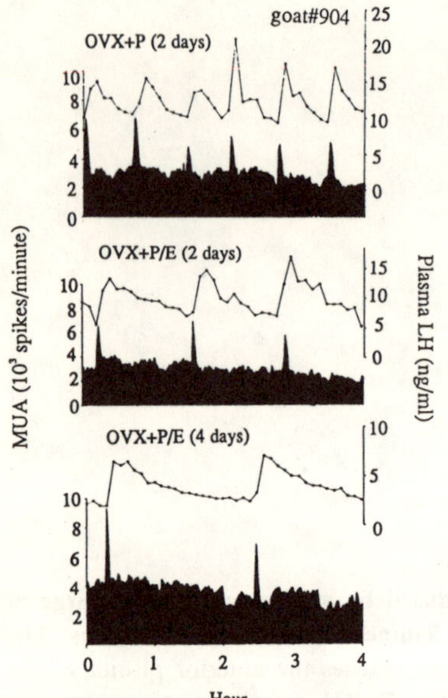

Fig. 5. Synchrony of LH pulses in the systemic circulation and the hypothalamic GnRH pulse generator activity as assessed by multiunit activity (MUA) in an ovariectomized (OVX) goat treated with progesterone (P) and/or estradiol (E). MUA was recorded under conscious and unrestrained conditions from electrodes placed in the medial basal hypothalamus. Each abrupt increase in MUA (the MUA volley) is followed by an LH pulse with 7–8 min latency regardless of inter-pulse intervals. It is therefore indicated that an MUA volley reflects an increased neural activity that is associated with the neurosecretion of GnRH from the nerve terminals. (From Mori *et al.*, 1991)

generator activity were first reported in ovariectomized rhesus monkeys (Knobil, 1981; Wilson *et al.*, 1984) and rats (Kawakami *et al.*, 1982). The technique is a useful tool for accessing directly the central component of the neuroendocrine system which regulates reproductive processes in these species (Kaufman *et al.*, 1985; Kesner *et al.*, 1986; Kimura *et al.*, 1991).

Similar attempts were made in sheep (Rasmussen & Malven, 1981; Thiery & Pelletier, 1981), but until quite recently conclusive results were not obtained in ruminants. Mori *et al.* (1991) reported the characteristic increases in neural activity coincident with the pulsatile release of LH in conscious unrestrained goats with chronically implanted electrodes in the medial basal hypothalamus (MBH). The recording electrodes were implanted bilaterally into the MBH, the arcuate-median eminence region, according to the stereotaxic procedure with radiographic monitoring (Mori *et al.*, 1990). A characteristic increase of the multiple unit activity (MUA volley) always preceded the LH pulse as shown in Fig. 5, and the temporal

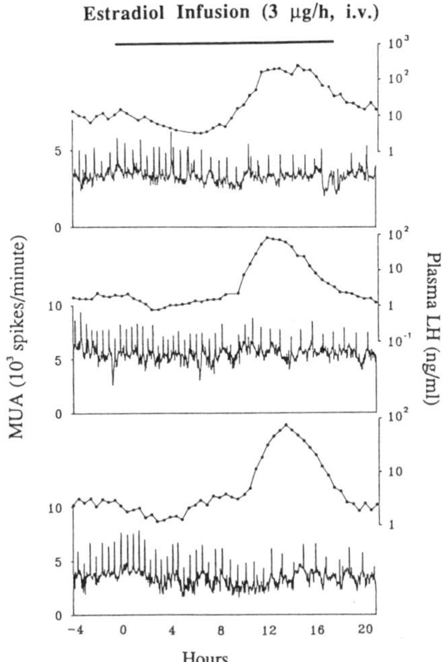

Fig. 6. Effects of intravenous infusion of estradiol (3 μg/hour for 16 hours) on LH secretion and the GnRH pulse generator activity in ovariectomized goats. The hypothalamic pulse generator activity as shown by the MUA volleys (see Fig. 5) continues even at the preovulatory-like LH surge induced by estradiol. (From Tanaka *et al.*, unpublished data)

correlation was consistent even when the pulse frequency was altered by the administration of an anesthetic or exogenous steroids. This experimental procedure has enabled a real-time recording of the GnRH pulse generator activity under conscious and unrestrained conditions for a considerably long period up to a few months if necessary. Another advantage of using this system is that the effects of various external and/or internal factors on the hypothalamic function can theoretically be discriminated from those on the pituitary gland as exemplified in Fig. 6. Here are shown changes in the hypothalamic MUA during the LH surge induced by estradiol in ovariectomized goats (Tanaka *et al.*, unpublished data). The MUA volleys continued with decreased frequency during the LH surge. It therefore seems unlikely at least under these experimental circumstances that LH pulses of very high frequency are released for the formation of LH surge. Caraty *et al.* (1989) found increased GnRH secretion in the pituitary portal circulation synchronized with LH surge in ovariectomized sheep given estradiol. As shown in Fig. 7 we could also demonstrate recently with a microdialysis

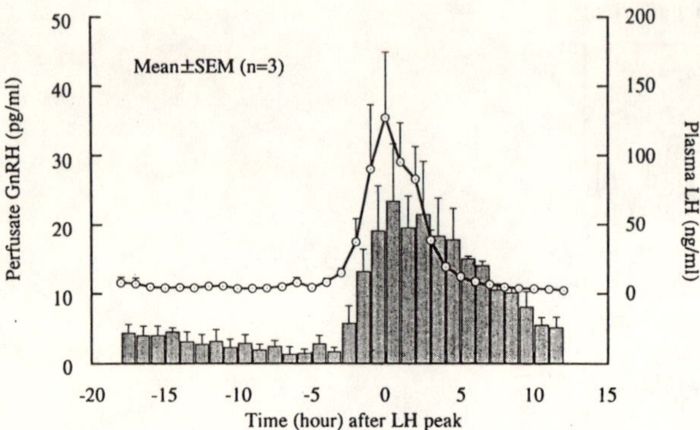

Fig. 7.   Changes in GnRH concentrations in microdialysis perfusate from the median eminence of ovariectomized goats during the LH surge induced by estradiol infused by the same protocol as in Fig. 6. The data from 3 goats are normalized to the time of LH peak. (From Manabe *et al.*, unpublished data)

TABLE 1
Distribution of immunoreactive GnRH cells (%) in the goat brain

Goat No.	(*n*)*	hDBB-MS	POA	AHA-LH	ARC-ME	others
934	(256)	7.0	56.2	23.0	9.0	4.8
943	(420)	10.2	59.8	21.2	6.9	1.9
952	(268)	4.1	66.1	20.1	7.8	1.9
959	(332)	5.7	60.6	25.3	6.0	2.4
961	(408)	3.2	70.1	16.9	5.9	3.9
mean		6.0	62.6	21.3	7.1	3.0

* Number of immunoreactive GnRH cells observed.
hDBB-MS, horizontal limb of diagonadal band of Broca-medial septum; POA, preoptic area; AHA-LH, anterior hypothalamic area-lateral hypothalamus; ARC-ME, arcuate-median eminence.

technique that GnRH secretion at the midian eminence increased during the LH surge induced by estradiol in ovariectomized goats (Manabe *et al.*, unpublished data). Herman & Adams (1990) reported that immunoneutralization against GnRH eliminated estradiol-induced LH surge in ovariectomized ewes. Taken together, the positive feedback effects of estradiol appear to be exerted through the neuronal circuit that is intrinsically different from the GnRH pulse generator dictating basal LH secretion.

The origin of pulsatile signals has not been determined yet. In the goat the vast majority of GnRH containing perikarya are immunohistochemically identified in the rostral portions of the hypothalamus with the highest concentrations in the medial preoptic area as shown in Table 1 (Hamada *et al.*, 1990). On the contrary, in the rhesus monkey, the major hypothalamic GnRH cell group resides in the periventricular as well as in the tuberal regions (Silverman *et al.*, 1982). Despite the apparent interspecies differences in the distribution of the hypothalamic GnRH cell bodies, MUA volleys associated with LH pulses were always recorded from the same region of the hypothalamus, *i.e.* the MBH. It is therefore suggested that MUA volleys are recorded from either the axonal tract or the terminal field of GnRH neurons but not from perikarya themselves. Alternatively, the recorded action potentials are originated from a neural oscillator which may lie in close proximity to the GnRH terminals. Definite conclusions must await more detailed electrophysiological as well as immunohisto-chemical studies.

GONADOTROPIN SECRETION AND PHOTOPERIOD

Effects of Photoperiod on Ovarian Function

Since the central mechanism regulating the annual reproductive cycles has elasticity, these cycles are entrainable to a shift of environmental seasonality just as the circadian system can soon be re-entrained to a shift of

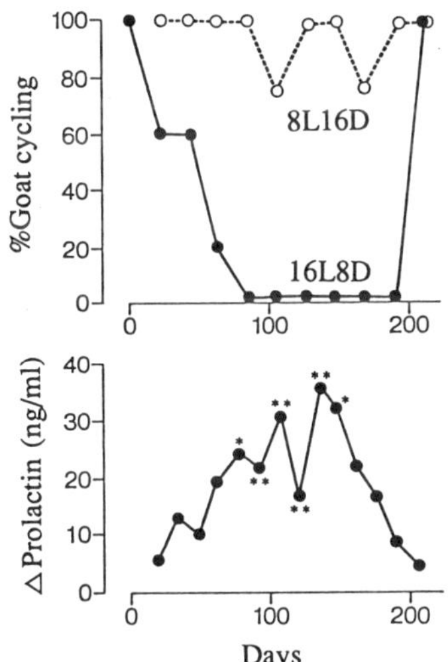

Fig. 8.   Photoperiodic control of ovulatory cyclicity and prolactin secretion in the goat. Two groups of normally cycling female goats were continuously kept under short days (8L16D) or transferred to long days (16L8D) on day 0. The percentage of animals recurring ovulatory cycles was assessed from plasma progesterone profiles. Plasma prolactin concentrations increased under long days, and the difference between the two groups ($\triangle$prolactin) became significant (*$P < 0.05$, **$P < 0.01$) during the period coinciding with the long-day induced anovulation. (From Mori *et al.*, 1984, 1985)

light-dark cycle. For example, it has long been known that the breeding season of sheep is inverted within a few years of a transfer from the north to the south hemisphere and vice versa, and the seasonality can also be controlled by artificial manipulation of photoperiod (Yeates, 1949).

Female goats show the ovulatory cycles under short days whereas they become anovulatory under long days suggesting that the photoperiod is of critical importance (Fig. 8.) (Mori *et al.*, 1984, 1985). The anti-gonadal effect of long days lasts only for several months, and then the goat becomes refractory to the long-day inhibition (photorefractoriness) and a gonadal recrudescence takes place. Both the gonadal suppression by long days and the subsequent resumption of ovarian cycles by photorefractoriness are considered to be essential for the institution of reproductive seasonality in this species. Prolactin secretion by the pituitary gland is stimulated under long days as shown in the lower panel of Fig. 8. The timing of hypersecretion of prolactin is temporally correlated with the anovulatory period. It is unknown whether there is any causative relationship between the hypogonadism and the hyperprolactinemia both induced by long days, but prolactin secretion has been shown to be one of the appropriate parameters by which to examine the effect of photoperiod on the reproductive endocrine system.

Photoperiodic Control of LH Secretion
To examine how the long-day photoperiod interferes with ovarian cyclicity the preovulatory endocrine sequence was compared under different photoperiodic conditions (Ohmori *et al.*, unpublished data). Under conditions of 10 h light and 14 h dark (10L14D) there was a typical preovulatory rise in plasma estradiol that was followed by the LH/FSH surges, and subsequent ovulation resulted in a normal luteal formation (Fig. 9, left panel). In contrast, under 16 h light and 8 h dark (16L8D) conditions neither an increase of estradiol secretion nor LH/FSH surge was observed. Post-luteolytic increase of LH pulse frequency was absent under the inhibitory photoperiod (Fig. 9, right panel). Since the acceleration of LH pulsatility is considered to be a prerequisite for normal follicular development (Karsch *et al.*, 1984; Martin, 1984), inadequate patterns of LH secretion would be responsible for the ovulatory failure under long days.

Photoperiodic control of LH secretion was also examined in a simpler experimental preparation (Mori *et al.*, 1987). Ovariectomized goats were subcutaneously implanted with an estradiol capsule to maintain physiological blood levels of estradiol. They were then exposed to long days (16L8D, from Day 0) or to short days (10L14D) to assess the effects of photoperiod on the sensitivity to the estradiol negative feedback. Exposure to the

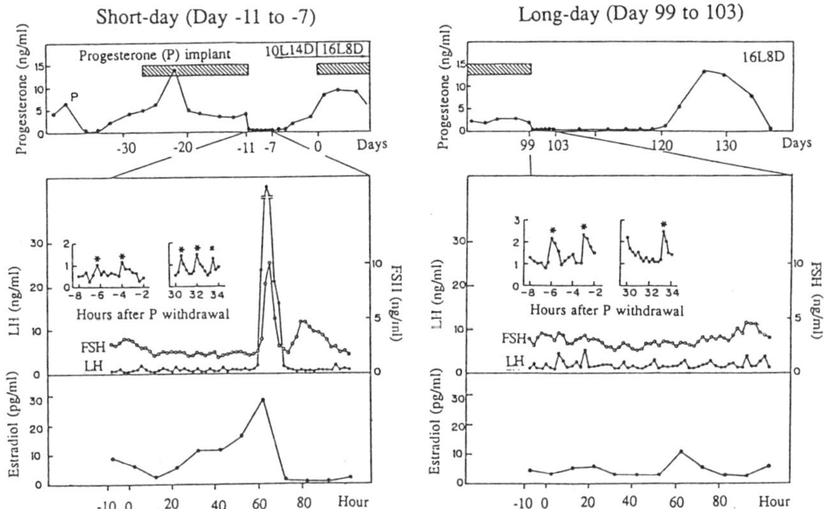

Fig. 9.   Effects of photoperiod on the preovulatory endocrine sequence in female goats. Under short days (left panel) following a removal of progesterone implants (0 hr) plasma estradiol rose and preovulatory LH/FSH surges were induced. In contrast, under long days (right panel) neither the estradiol rise nor LH/FSH surges were observed. Note the absence under long days of increased LH pulse frequency following progesterone withdrawal that was seen under short days, suggesting the role of GnRH pulse generator activity in the photoperiodic control of ovarian function in this species. (From Ohmori *et al.*, unpublished data)

TABLE 2

Plasma concentrations of LH and estradiol in ovariectomized estradiol-implanted goats kept under a 10L14D (Group 1, $n=4$) or 16L8D (Group 2, $n=4$) regimen

Day[†]	LH (ng/ml)		Estradiol (pg/ml)	
	Group 1	Group 2	Group 1	Group 2
0–40	4.9±0.3	5.5±0.6	10.0±1.8	8.5±2.3
41–80	5.1±0.5	3.2±0.6*	7.0±1.8	5.0±1.6
81–120	3.7±0.5	1.8±0.3**	4.0±1.0	5.3±1.8
121–160	3.5±0.7	1.4±0.2*	6.0±0.2	8.0±2.0
161–200	3.0±0.6	1.3±0.1*	8.0±2.5	9.8±2.1
201–260	3.7±0.4	3.3±0.8	4.5±1.0	7.0±1.8

Values are mean±SEM of 12–20 pooled observations.
[†] Group 2 goats were transferred from 10L14D to 16L8D on Day 0.
*$P<0.05$, **$P<0.01$ compared with Group 1.

long-day photoperiod lowered plasma LH concentrations in the estradiol-treated ovariectomized goats with a time course similar to the long-day induced anovulation in ovary-intact goats, suggesting an enhanced negative

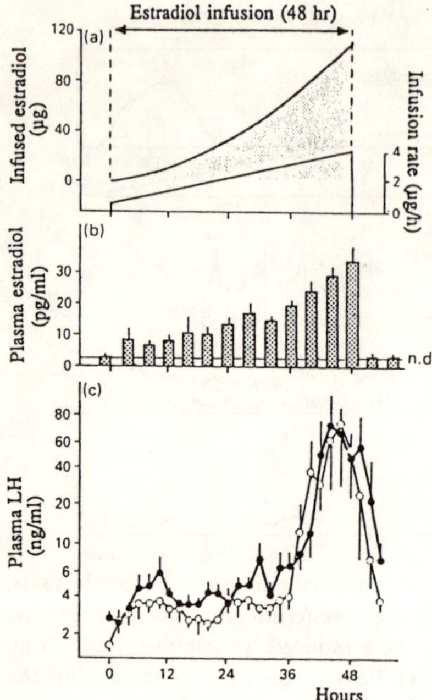

Fig. 10.   Outline of the programmed administration of estradiol. (a) The infusion rate was linearly increased from 0.5 to 4.1 μg/h, and the accumulated amount of infused estradiol during the 48 h period averaged 110.4 μg. (b) Plasma estradiol concentration showed a gradual rise within the physiological range as infusion rate increased. (c) Plasma LH profiles during the estradiol infusion on Day 30 of 16L8D (closed circles) and 8L16D (open circles) regimen. Each point represents the mean value ($\pm$SEM) for 4 goats. (From Mori *et al.*, 1987)

feedback effect of estradiol on basal LH secretion under a long-day regimen (Table 2). The seasonal change of estradiol negative feedback was first reported in sheep by Legan *et al.* (1977), and the importance of LH pulse frequency underlying this seasonal change was later demonstrated by Goodman *et al.* (1982).

On days 30, 60, 100, 149, and 279 an LH surge was induced by programmed infusion of estradiol that mimicked a preovulatory increase of circulating estradiol (Fig. 10). The positive feedback effect of estradiol on LH secretion was also found to be enhanced under long days, since the latency from the onset of estradiol infusion to the LH surge became shorter and therefore less estradiol was required for induction of the LH surge (Fig. 11). These results might indicate that the hypothalamo-pituitary axis of the goat becomes hypersensitive to the positive as well as the negative feedback effect of estradiol under long days. The ovaries of anestrous ewes and goats

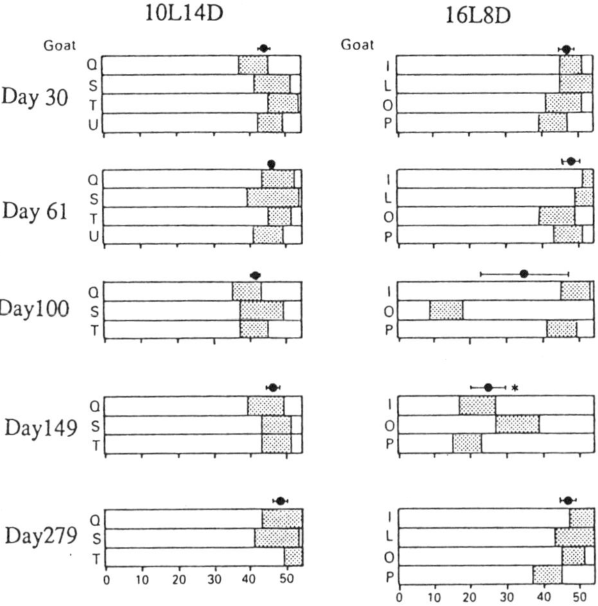

Hour from start of estradiol infusion

Fig. 11.   Timings of induced LH surge (shaded area) in individual ovariectomized goats kept under short days (10L14D) or long days (16L8D). Mean ($\pm$ SEM) time of maximum LH concentration for the two groups is presented at the top of each panel. *$P < 0.05$ between groups on the same trial day. (From Mori *et al.*, 1987)

lack corpora lutea but contain a considerable number of antral follicles (Yoshioka, 1961) which secrete estradiol in response to a pulsatile discharge of LH (Scaramuzzi & Baird, 1977). Provided the hypothalamic-pituitary axis were to become more susceptible to estradiol at this time, even a small amount of estradiol released from such follicles might be able to trigger a premature LH surge. Since it is known that premature exposure to the LH surge renders developing follicles atretic (Harman *et al.*, 1975), undesirable ovulation during the anestrous season could thus be avoided. Photoperiodic control of LH secretion may be exerted by both the indirect way *via* steroidal feedback and the direct action on the GnRH pulse generator activity. As a novel experimental tool for assessing these actions, the real-time monitoring system of electrophysiological manifestation of GnRH pulse generator activity described above is expected to be useful.

ROLE OF THE PINEAL GLAND IN THE REPRODUCTIVE PHOTOPERIODICITY

In the photoperiodic control of reproduction it is now generally accepted that the secretory pattern of melatonin by the pineal gland is of decisive importance in conveying information on changing daylength to the neuro-endocrine system which eventually accelerates or suppresses the GnRH pulse generator activity. Herein described is evidence supporting the current hypothesis that the brain responds to changes in the duration of the daily melatonin signal to entrain reproductive seasonality.

Diurnal Patterns of Melatonin Secretion

Melatonin secretion by the pineal gland is restricted to the night, and there is a clear 24 hour rhythmicity in the plasma level of melatonin (Maeda *et al.*, 1984). Similar diurnal change of melatonin was found in the cerebrospinal fluid (CSF) sequentially collected from the lateral ventricle of conscious goats as shown in Fig. 12 (Kanematsu *et al.*, 1989). The magnitude of nocturnal melatonin elevation is, however, much bigger in CSF than in peripheral circulation, suggesting the possibility that pineal melatonin reaches its putative target site, the hypothalamus, directly via the brain ventricular system. The night-interruption by exposure to the light results in an abrupt cessation of melatonin secretion, and it is known that

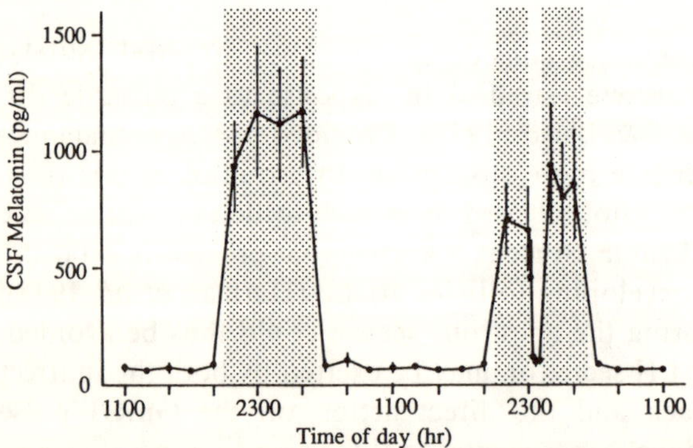

Fig. 12.   Concentrations of melatonin in cerebrospinal fluid (CSF) collected from the lateral ventricle of a conscious goat. High concentrations of CSF melatonin are restricted to the dark phase (shaded area) showing a distinct 24 hour rhythmicity of melatonin secretion by the pineal gland. Each point represents mean value ($\pm$SEM) for 6 animals. One hour exposure to the light on the subsequent night resulted in an abrupt cessation of melatonin secretion. (From Kanematsu *et al.*, 1989)

the light intensity required for this melatonin suppression varies from species to species in a range between 0.02 lux in the cotton rat to 1,000 lux or greater in man (see Nozaki *et al.*, 1990).

Removal of Melatonin Signal

The effects of the abolition of the diurnal change in melatonin secretion on the photoperiodic response of the gonadal axis and prolactin secretion to long days were examined in female goats (Maeda *et al.*, 1986). Female goats

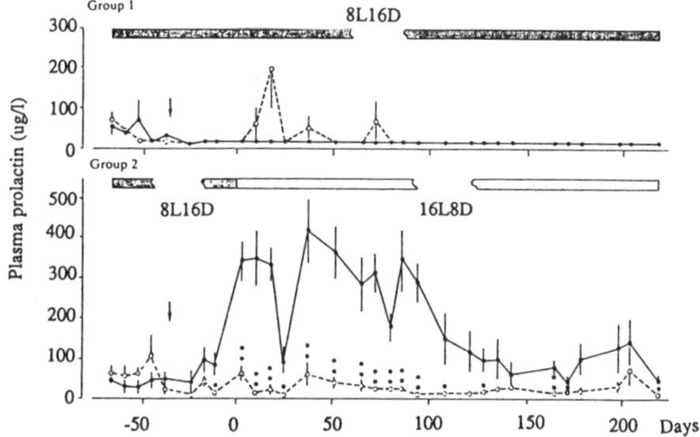

Fig. 13. Involvement of the pineal gland in the photoperiodic control of prolactin secretion in goats. Pineal denervation (arrow) resulted in an elimination of plasma prolactin rise that was seen in pineal intact animals (closed circles) following the transfer from short days to long days. (From Maeda *et al.*, 1986)

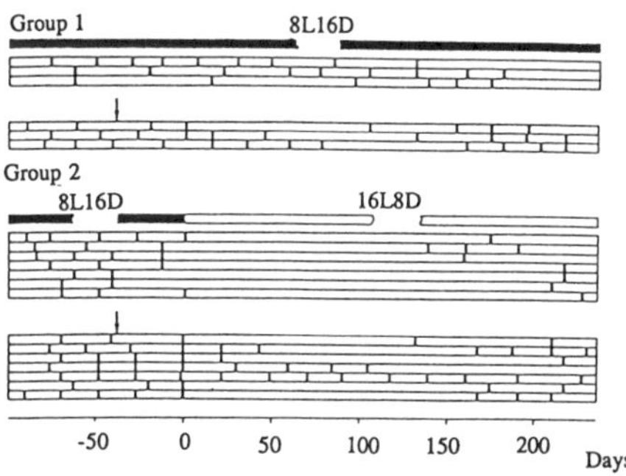

Fig. 14. Effects of pineal denervation on the ovarian function of female goats under short days (8L16D) and long days (16L8D). Ovulations were estimated by fluctuations of plasma progesterone levels (solid vertical bars) in individual goats as shown by horizontal columns. Arrows indicate the day of pineal denervation. In the pineal intact goats an exposure to long days resulted in a cessation of ovulatory cyclicity. This response was, however, blocked or largely attenuated in the pineal denervated animals. (From Maeda *et al.*, 1986)

reared under short days (8L16D) were bilaterally superior cervical gan-
glionectomized (SCGX). One month after surgery, both SCGX and intact
control goats were divided into two groups and exposed either to short days
or to long days (16L8D). Under short days both SCGX and intact animals
ovulated periodically and basal plasma prolactin levels were maintained
(Figs. 13 and 14). In intact controls, exposure to long days suppressed
ovulation and increased prolactin secretion for the first 150–200 days of
exposure, thereafter the animals became photorefractory, ovulation recur-
red and prolactin secretion returned to basal levels. Superior cervical
ganglionectomy abolished or weakened the suppression of gonadal function and eliminated the increase in prolactin secretion induced by exposure
to long days in the intact control.

These results suggest that the pattern of melatonin secretion under long
days is necessary to induce the photoperiodic response of gonadal function
and prolactin secretion to long days.

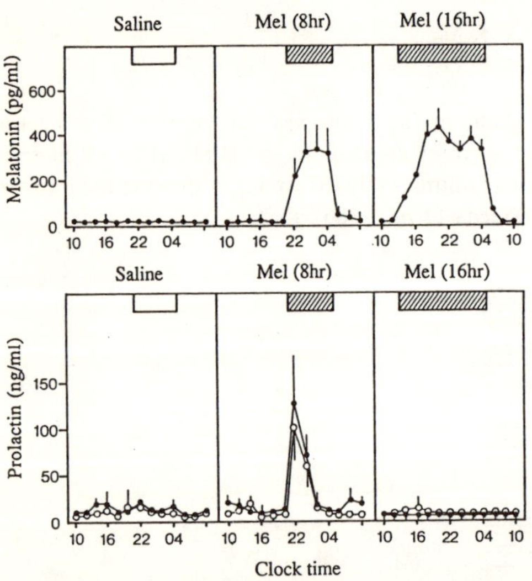

Fig. 15.   Restoration of 24 hour rhythms of circulating melatonin mimicks the effects of
photoperiodic change on prolactin secretion. Daily s.c. infusion of melatonin for 8 hours
reproduced a long day pattern of melatonin secretion and stimulated prolactin secretion
as a long-day treatment did in intact animals (middle panels), whereas the simulation of
a short-day pattern with 16 hour infusion of melatonin suppressed prolactin secretion as
seen under short days. The data are mean ($\pm$SEM) of 5 animals obtained under 16L8D
(closed circles) and 8L16D (open circles) showing that environmental photoperiod had
no appreciable influence on the melatonin action. (From Mori & Okamura, 1986)

Replacement of Circulating Melatonin Rhythms

Melatonin signal was restored in the pineal denervated goat by the timed melatonin infusion, and effects on prolactin secretion were examined (Mori & Okamura, 1986). In SCGX goats, which had failed to coordinate their prolactin secretion with the prevailing photoperiod, melatonin was subcutaneously infused (20 μg/h) daily for 8 h (the long-day-type infusion) or for 16 h (the short-day-type infusion) to mimic the nocturnal profile of plasma melatonin under long days or short days, respectively (Fig. 15). After several days the long-day-type melatonin infusion accelerated prolactin secretion, inducing a nocturnal rise in plasma prolactin; this was comparable to that seen in the pineal intact goats under the long photoperiods. On the other hand, the short-day-type melatonin infusion suppressed prolactin secretion throughout a day as the short-day treatment did in intact goats. The prevailing photoperiod appeared to have no distinct effect on these prolactin responses to exogenous melatonin, which were indistinguishable under 16L8D and 8L16D conditions. The results indicate that the information about external light-dark cycles is converted by the pineal gland into the endocrine signal as a daily pattern of melatonin secretion, which eventually regulates prolactin secretion from the pituitary gland of the goat. It is therefore suggested that in the photoperiodic control of

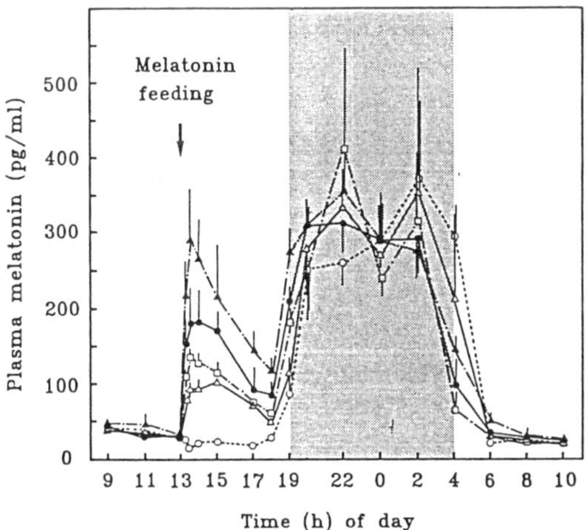

Fig. 16. Diurnal changes in plasma melatonin concentrations in ewes fed varying doses of melatonin in the early afternoon (arrow) under the natural long-day photoperiod. Daily doses of melatonin were 0 mg (○), 1 mg (△), 2 mg (□), 3 mg (●) and 4 mg (▲). Each point represents the mean (±SEM) of 4 ewes. The shaded area indicates the period of darkness. (From Kusakari *et al.*, 1991)

prolactin secretion the pineal gland is acting as an interface which transduces the neural input from the retina to the endocrine output, melatonin. Bittman *et al.* (1983a, b) have clearly demonstrated that this is also the case for LH release.

Melatonin Induced Out-of-seasonal Reproduction

Based on the above mentioned idea the manipulation of diurnal rhythm of circulating melatonin has been examined as an alternative means of induction of out-of-seasonal reproduction in domestic ruminants (Arendt *et al.*, 1983; Kennaway *et al.*, 1982; Nett & Niswender, 1982). It has been known for many years that the short-day treatment can induce earlier onset of the breeding season in ewes and goats (Yeates, 1949; Yoshioka, 1961). Instead of placing the animal in a dark room melatonin was administered to anestrous ewes by various means; for example, it was adsorbed onto pelleted food and administered orally at a fixed time of the light phase as shown in Fig. 16 (Kusakari *et al.*, 1991). Plasma melatonin increased in a dose-dependent manner, and the magnitude of rise was similar to endogenous nocturnal peaks in sheep given 4 mg of melatonin. If this tentative melatonin rise is perceived by the animal as a signal of night onset, daily melatonin feeding should have the same effect as short-day treatment and will induce an ovulatory cyclicity under long days. In fact, the

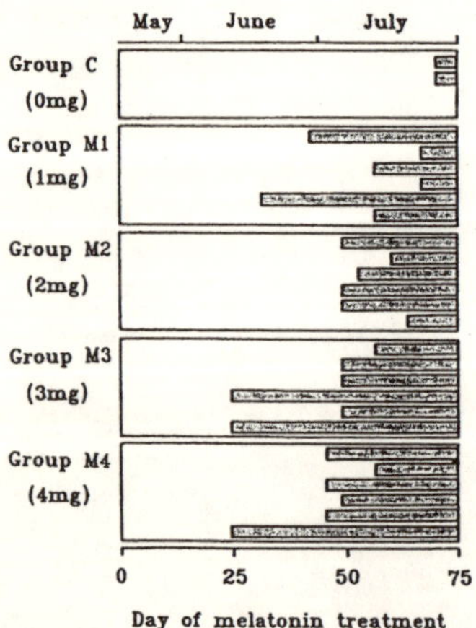

Fig. 17. Occurrence of cyclical ovarian activity (solid bars) in individual ewes given varying daily doses of melatonin. Reproductive status was assessed from plasma progesterone profiles. The intervals from the start of melatonin treatment averaged 72.5, 53.0, 53.6, 42.0 and 44.3 days for the groups receiving daily 0, 1, 2, 3, and 4 mg melatonin, respectively. (From Kusakari *et al.*, 1991)

melatonin treated animals showed an earlier onset of the breeding season
(Fig. 17).

In addition to the timed melatonin administration it has been demonstrated that a subcutaneous melatonin implantation which maintains high
plasma melatonin levels throughout a day is also effective for inducing an
out-of-seasonal reproduction in ewes (Mori *et al.*, 1987, 1990). Continuous
high melatonin levels may have masked the endogenous pattern of
melatonin secretion and thus released the animal from the long-day suppression. Various methods have now become available for the chronic
administration of melatonin including subcutaneous capsules and intra-
rumen pellets (Poulton *et al.*, 1986).

Mechanism of Melatonin Action

The role of melatonin in the photoperiodic responses is thus well established. The most important question, therefore, is where and how melatonin acts. Single intravenous injection of melatonin (5 μg/head) causes an

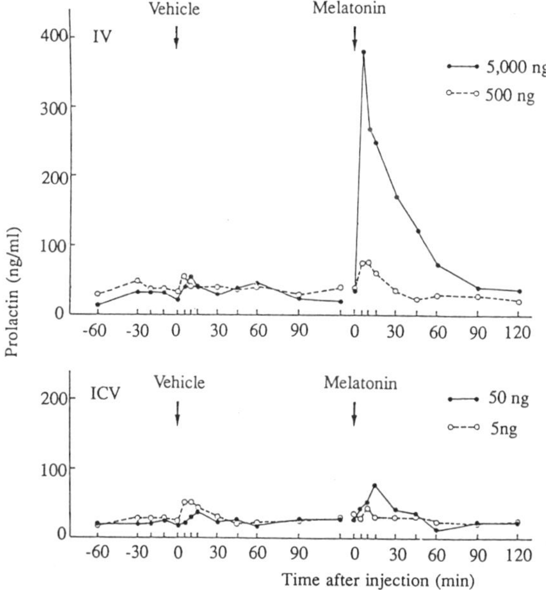

Fig. 18. Acute effects on prolactin secretion of melatonin given systemically (IV) or
intracerebroventricularly (ICV) to the ovariectomized, estradiol-treated (carrying a
subcutaneous capsule) goat. Melatonin was dissolved in 0.25–0.5% EtOH saline solution
(vehicle) and administered through a jugular catheter (IV) or injection cannulae chronically implanted in the lateral ventricles. (From Takeuchi *et al.*, unpublished data)

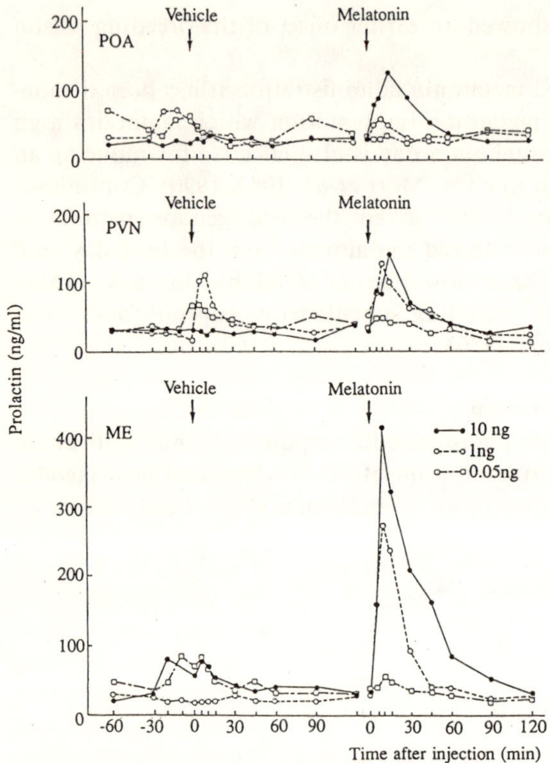

Fig. 19. Acute effects of hypothalamic microinjections of melatonin on prolactin secretion in the ovariectomized, estradiol-treated goat. Melatonin (0.05–10 ng) was dissolved in 20 μl of 0.05% EtOH saline solution and slowly (20 sec) administered through pre-implanted cannulae into the pre-optic area (POA), the paraventricular nucleus (PVN) and the median eminence (ME) of the hypothalamus. Note that much smaller amount of melatonin (about one 500 th of systemic administration) was effective in eliciting acute prolactin secretion in a dose-dependent manner when administered into ME, but not POA or PVN. (From Takeuchi *et al.*, unpublished data)

acute and temporal rise of prolactin secretion in the ovariectomized, estradiol-implanted goats (Fig. 18, Takeuchi *et al.*, unpublished data). The effects of microinjections of melatonin (0.05–10 ng) into hypothalamic nuclei were then examined. Much smaller amounts of melatonin, as compared with systemic administration, stimulated prolactin secretion in a dose dependent manner when administered to the median eminence (ME) (Fig. 19). Microinjections to neither the paraventricular nucleus (PVN) nor the preoptic area (POA) produced such a clear response as did to ME. Similarly, melatonin implanted in the MBH, but not in other areas of the hypothalamus (*e.g.*, POA) caused a premature increase of LH secretion in ovariectomized, estradiol-implanted ewes (Malpaux *et al.*, 1990) and accelerated the gonadotropin secretion and testicular development in rams (Lincoln & Maeda, 1992) kept under the long day condition. Melatonin is postulated to act through specific membrane bound, high affinity receptors to induce its biological effects (Morgan & Williams, 1989), and these receptors are most frequently distributed in the pars tuberalis of the pituitary gland that is adjacent to the median eminence in sheep (Morgan *et al.*, 1989; Pelletier *et al.*, 1990). Melatonin presumably acts on GnRH secretion indirectly within or close to the MBH by affecting the activity of either the catecholaminergic or the peptidergic neurons normally regulating GnRH secretion, or alternatively by the paracrine mode of action at pars tuberalis. The intracerebral target site and the cellular mechanism for the melatonin action are the most interesting and urgent subjects in the future study of the central mechanism of reproductive photoperiodicity in mammalian species, as suggested by Stankov & Reiter (1990).

CONCLUSION

Seasonal reproduction is an annual rhythm of gonadal activity that is widely seen among mammalian species inhabiting areas extending from temperate zones to higher latitudes. Environmental photoperiod has a profound effect on the reproductive endocrine function in the seasonal breeders; *e.g.* the transfer of female goats from the short-day to the long-day photoperiod results in a cessation of ovulatory cyclicity. In the first section of this article the secretory patterns of LH and the role of the hypothalamic GnRH pulse generator during the normal ovulatory cycle in the goat were discussed. In the second part the photoperiodic modification of the reproductive endocrine system was described. Evidence suggests that under a suppressive photoperiod there is a reduction in the frequency of intermittent discharge of hypothalamic GnRH into the pituitary portal circulation that consequently lowers circulating levels of gonadotropins and results in gonadal regression. In the third section, the pathway of the photic information was considered. As in many other mammalian species investigated plasma melatonin levels in the goat show a distinct daily rhythm with the secretory phase restricted to the dark period. Denervation of the pineal gland results in an elimination of this endogenous 24 hour rhythmicity of melatonin secretion. Pineal denervated goats are no longer able to adjust their endocrine state to the prevailing photoperiod. Restoration of plasma melatonin rhythm of either the long-day or the short-day type by timed melatonin infusion induces endocrine responses similar to those photoperi-

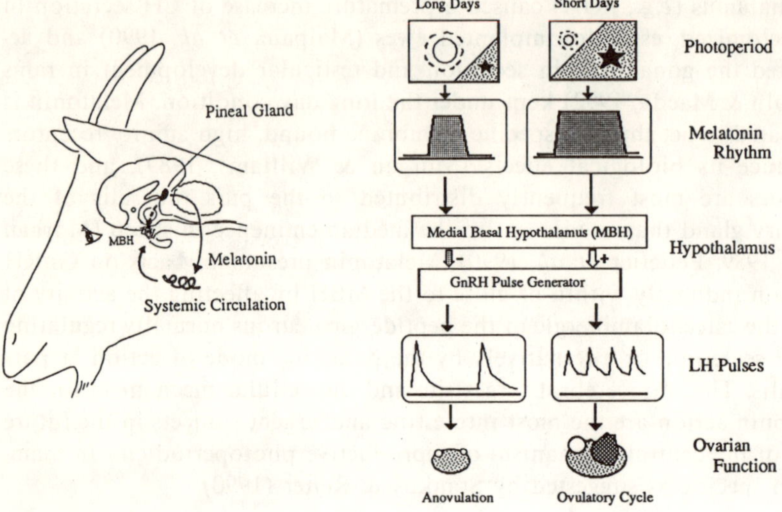

Fig. 20. Hypothetical neuroendocrine mechanism underlying the photoperiodic modification of gonadal function in short-day breeders. Information on changing daylength is mediated by the pineal gland through its secretory pattern of melatonin. Melatonin reaches its postulated target site, the medial basal hypothalamus (MBH), either directly *via* cerebroventricular system or indirectly *via* the systemic circulation through the blood-brain barrier. The melatonin signal mediating inhibitory photoperiod (*e.g.*, long daylength for the short-day breeders) influences either catecholaminergic or peptidergic neurons and slows down the hypothalamic GnRH pulse generator activity, which consequently suppresses pituitary gonadotropin secretion and induces gonadal regression.

odically evoked in control animals. In fact the reproductive seasonality can be manipulated by the melatonin treatment. Earlier onset of breeding season was induced in seasonally anestrous ewes subjected either to daily timed melatonin administration that mimicked short-day profiles of plasma melatonin or to continuous treatment with s.c. implants that blunted the endogenous melatonin rhythm of long days. Melatonin is therefore an endocrine signal conveying information of night length, and the changing photoperiod is transduced by the pineal gland into the secretory profile of melatonin that modifies the GnRH pulse generator activity and, consequently, the pattern of gonadotropin secretion (Fig. 20).

ACKNOWLEDGEMENTS

Most of the work herein described was performed in the Laboratory of Veterinary Reproduction, Tokyo University of Agriculture and Technology, during the past ten years while the author was enrolled. I would therefore like to thank Professor K. Hoshino and all the graduate and undergraduate students of the laboratory for their continuous help, enthusiasm and encouragement. I am also grateful to my collaborators from outside the campus, especially to Drs. S. Hayashi, M. Ichikawa, Y. Kanai, Y. Kano, N. Kusakari, K.-I. Maeda, M. Nishihara, M. Nozaki and M. Takahashi for their consistent support, discussion and criticism.

REFERENCES

Arendt, J., Symons, A.M., Laud, C.A. & Pryde, S.J. (1983). Melatonin can induce early onset of the breeding season in ewes. *Journal of Endocrinology* **97**, 395–400.

Bittman, E.L., Dempsey, R.J. & Karsch, F.J. (1983a). Pineal melatonin secretion drives the reproductive response to daylength in the ewe. *Endocrinology* **113**, 2276–2283.

Bittman, E.L., Karsch, F.J. & Hopkins, J.W. (1983b). Role of the pineal gland in ovine photoperiodism: regulation of seasonal breeding and negative feedback effects of estradiol upon luteinizing hormone secretion. *Endocrinology* **113**, 329–396.

Bronson, F.H. (1988). Seasonal regulation of reproduction in mammals. In: *The Physiology of Reproduction* (Knobil, E. and Neil, J., eds.), Raven Press, New York, pp. 1831–1871.

Caraty, A., Locatelli, A. & Martin, G.B. (1989). Biphasic secretion of gonadotropin-releasing hormone (GnRH) in ovariectomized ewes injected with oestradiol. *Journal of Endocrinology* **123**, 375–382.

Clarke, I.J. & Cummins, J.T. (1985). The temporal relationship between gonadotropin releasing hormone (GnRH) and luteinizing hormone (LH) secretion in ovariectomized ewes. *Endocrinology* **111**, 1737–1739.

Desjardins, C. & Lopez, M.J. (1980). Sensory and nonsensory modulation of testis function. In: *Testicular Development, Structure and Function* (Steinberger, A. and Steinberger, E., eds.), Raven Press, New York, pp. 381–388.

Foster, D.L., Karsch, F.J., Olster, D.H., Ryan, K.D. & Yellon, S.M. (1986). Determinants of puberty in seasonal breeder. *Recent Progress in Hormone Research* **42**, 330–384.

Goodman, R.L., Bittman, E.L., Foster, D.L. & Karsch, F.J. (1982). Alterations in the control of luteinizing hormone pulse frequency underlie the seasonal variation in estradiol negative feedback in the ewe. *Biology of Reproduction* **27**, 580–589.

Hamada, T., Shimizu, T., Mori, Y. & Ichikawa, M. (1990). Immunohistochemical study on GnRH neurons in the Shiba goat. *Proceedings of the Japan Society for Comparative Endocrinology* **5**, 24.

Harman, S.M., Louvet, J.P. & Ross, G.T. (1975). Introduction of estrogen and gonadotropins on follicular atresia. *Endocrinology* **96**, 1145–1152.

Herman, M.E. & Adams, T.E. (1990). Gonadotropin secretion in ovariectomized ewes: effect of passive immunization against gonadotropin-releasing hormone (GnRH) and infusion of a GnRH agonist and estradiol. *Biology of Reproduction* **42**, 273–280.

Kanematsu, N., Mori, Y., Hayashi, S. & Hoshino, K. (1989). Presence of a distinct 24 hour melatonin rhythm in the ventricular cerebrospinal fluid of the goat. *Journal of Pineal Research* **7**, 143–152.

Karsch, F.J., Bittman, E.L., Foster, D.L., Goodman, R.L., Legan, S.J. & Robinson, J.E. (1984). Neuroendocrine basis of seasonal reproduction. *Recent Progress in Hormone Research* **40**, 185–225.

Kaufman, J.M., Kesner, J.S., Wilson, R.C. & Knobil, E (1985). Electrophysiological manifestation of luteinizing hormone-releasing hormone pulse generator activity in the rhesus monkey: Influence of α-adrenergic and dopaminergic blocking agents. *Endocrinology* **116**, 1327–1333.

Kawakami, M., Uemura, T. & Hayashi, R. (1982). Electrophysiological correlates of pulsatile gonadotropin release in rats. *Neuroendocrinology* **35**, 63–67.

Kesner, J.S., Kaufman, J.M., Wilson, R.C., Kuroda, G. & Knobil, E. (1986). On the short-loop feedback of the hypothalamic luteinizing hormone releasing hormone 'pulse generator' in the rhesus monkey. *Neuroendocrinology* **42**, 109–111.

Kennaway, D.J., Gilmore, T.A. & Seamark, R.F. (1982). Effects of melatonin implants on the circadian rhythm of plasma melatonin and prolactin in sheep. *Endocrinology* **110**, 2186–2188.

Kimura, F., Nishihara, M., Hiruma, H. & Funabashi, T. (1991). Naloxone increases the frequency of the electrical activity of luteinizing hormone-releasing hormone pulse generator activity in long-term ovariectomized rats. *Neuroendocrinology* **53**, 97–102.

Knobil, E. (1980). The neuroendocrine control of the menstrual cycle. *Recent Progress in Hormone Research* **36**, 53–88.

Knobil, E. (1981). Patterns of hypothalamic signals and gonadotropin secretion in the rhesus monkey. *Biology of Reproduction* **24**, 44–49.

Kusakari, N., Tajima, Y., Itoh, S., Senna, K., Serikawa, S., Hatta, T., Ohara, M. & Mori, Y. (1991). Diurnal changes in plasma melatonin and the timing of reproductive onset in anestrous sheep daily fed melatonin. *Journal of Veterinary Medical Science* **53**, 457–461.

Legan, S.J., Karsch, F.J. & Foster, D.L. (1977). The endocrine control of seasonal reproductive function in the ewe: a marked change in response to the negative feedback action of estradiol on luteinizing hormone secretion. *Endocrinology* **101**, 818–824.

Levin, J.E., Pau, K.Y.F., Ramirez, V.D., Jackson, G.L. (1982). Simultaneous measurement of luteinizing hormone-releasing hormone and luteinizing hormone release in unanesthesized, ovariectomized sheep. *Endocrinology* **111**, 1449–1455.

Lincoln, D.W., Fraser, H.M., Lincoln, G.A., Martin, G.B. & MacNeilly, A.S. (1985). Hypothalamic pulse generators. *Recent Progress in Hormone Research* **41**, 369–419.

Lincoln, G.A. & Maeda, K.-I. (1992). Reproductive effects of placing micro-implants of melatonin in the mediobasal hypothalamus and preoptic area in rams. *Journal of Endocrinology* **132**, in press.

Lincoln, G.A. & Short, R.V. (1980). Seasonal breeding: nature's contraceptive. *Recent Progress in Hormone Research* **36**, 1–43.

Maeda, K.-I., Mori, Y. & Kano, Y. (1984). Diurnal changes in peripheral melatonin concentration in goats and effects of light or dark interruption. *Japanese Journal of Veterinary Science* **46**, 837–842.

Maeda, K.-I., Mori, Y. & Kano, Y. (1986). Superior cervical ganglionectomy prevents gonadal regression and increased plasma prolactin concentrations induced by long days in goats. *Journal of Endocrinology* **110**, 137-144.

Maeda, K.-I., Mori, Y. & Kano, Y. (1988). Involvement of melatonin in seasonal changes of the gonadal function and prolactin secretion in female goats. *Reproduction, Nutrition, Development* **28**, 487-497.

Malpaux, B.D.A., Gayrand, V., Maurice, F. & Thiery, J.C. (1990). Melatonin acts in the medial basal hypothalamus to control reproduction in the ewe. *Society for Research on Biological Rhythms 2nd Meeting*, Jacksonville, Florida, Abstract, 141.

Martin, G.B. (1984). Factors affecting the secretion of luteinizing hormone in the ewe. *Biological Review* **59**, 1-87.

Morgan, P.J. & Williams, L.M. (1989). Central melatonin receptors: Implications for a mode of action. *Experimentia* **45**, 955-965.

Morgan, P.J., Williams, L.M., Davidson, G., Lawson, W. & Howell, E. (1989). Melatonin receptors on ovine pars tuberalis: characterization and autoradiographic localization. *Journal of Neuroendocrinology* **1**, 1-4.

Mori, Y. & Kano, Y. (1984). Changes in plasma concentrations of LH, progesterone, oestradiol in relation to the occurrence of luteolysis, oestrus and time of ovulation in the Shiba goat (*Capra hircus*). *Journal of Reproduction and Fertility* **72**, 223-230.

Mori, Y. & Maeda, K.-I. (1991). Photoperiodism in mammals. In: *Handbook of Chronobiology* (Chiba, Y. and Takahashi, K., eds), Asakura, Tokyo, pp. 244-255 (in Japanese).

Mori, Y., Maeda, K. & Kano, Y. (1985). Photoperiodic control of prolactin secretion in the goat. *Japanese Journal of Animal Reproduction* **31**, 9-15.

Mori, Y., Maeda, K., Sawasaki, T. & Kano, Y. (1984). Effects of long days and short days on estrous cyclicity in two breeds of goats with different seasonality. *Japanese Journal of Animal Reproduction* **30**, 239-245.

Mori, Y., Nishihara, M., Tanaka, T., Shimizu, T., Yamaguchi, M., Takeuchi, Y. & Hoshino, K. (1991). Chronic recording of electrophysiological manifestation of the hypothalamic gonadotropin-releasing hormone (GnRH) pulse generator activity in the goat. *Neuroendocrinology* **53**, 392-395.

Mori, Y. & Okamura, H. (1986). Effects of timed melatonin infusion on prolactin secretion in pineal denervated goats. *Journal of Pineal Research* **3**, 77-86.

Mori, Y., Shimizu, K. & Hoshino, K. (1987). A rise in peripheral melatonin induces ovarian activity in anestrous sheep. *Japanese Journal of Animal Reproduction* **33**, 113-117.

Mori, Y., Shimizu, K. & Hoshino, K. (1990) Melatonin but not the ram effect reactivates quiescent ovarian activity of mid-anestrous ewes. *Japanese Journal of Veterinary Science* **52**, 773-779.

Mori, Y., Takedomi, T., Mizomoto, Y. & Hoshino, K. (1987). Induction of preovulatory endocrine events by programmed administration of progesterone and estradiol in ovariectomized goats. *Japanese Journal of Animal Reproduction* **33**, 36-40.

Mori, Y., Takeuchi, Y., Shimada, M., Hayashi, S. & Hoshino, K. (1990). Stereotaxic approach to hypothalamic nuclei of the Shiba goat with radiographic monitoring. *Japanese Journal of Veterinary Science* **52**, 339-349.

Mori, Y., Tanaka, M., Maeda, K., Hoshino, K. & Kano, Y. (1981). Photoperiodic modification of negative and positive feedback effects of oestradiol on LH secretion in ovariectomized goats. *Journal of Reproduction and Fertility* **80**, 523-529.

Nett, T.M. & Niswender, G.D. (1982). Influence of exogenous melatonin on seasonality of reproduction in sheep. *Theriogenology* **17**, 645–653.

Nozaki, M., Tsushima, M. & Mori, Y. (1990). Diurnal changes in serum melatonin concentrations under outdoor and indoor environment and light suppression of nighttime melatonin secretion in the female Japanese monkey. *Journal of Pineal Research* **9**, 221–230.

Pelletier, J., Castro, B., Roblot, G., Wylde, R. & de Revies, M.M. (1990). Characterization of melatonin receptors in the ram pars tuberalis: influence of light. *Acta Endocrinologica* **123**, 557-562.

Poulton, A.L., English, J., Symons, A.M. & Arendt, J. (1987). Changes in plasma concentrations of FSH and prolactin in ewes receiving melatonin and short photoperiod to induce early onset of breeding activity. *Journal of Endocrinology* **112**, 103–111.

Rasmussen, D.D. & Malven, P.V. (1981). Chronic recording of multiple-unit activity from the brain of conscious sheep. *Brain Research Bulletin* **7**, 163–167.

Scaramuzzi, R.J. & Baird, D.T. (1977). Pulsatile release of luteinizing hormone and the secretion of ovarian steroids in sheep during anestrus. *Endocrinology* **101**, 1801–1806.

Silverman, A.J., Antunes, J.L., Abrams, G.M., Nilaver, G., Thau, R., Robinson, J.A., Ferin, M. & Krey, L.C. (1982). The luteinizing hormone-releasing hormone pathways in rhesus and pigtailed monkeys: New observations on thick, unembedded sections. *Journal of Comparative Neurology* **211**, 309–317.

Stankov, B. & Reiter, R.J. (1990). Melatonin receptor: current status, facts, and hypotheses. *Life Science* **46**, 971–982.

Thiery, J.C. & Martin, G.B. (1991). Neurophysiological control of the secretion of gonadotropin-releasing hormone and luteinizing hormone in the sheep—a review. *Reproduction, Fertility and Development* **3**, 137–173.

Thiery, J.C. & Pelletier, J. (1981). Multiunit activity in the anterior median eminence and adjacent areas of the hypothalamus of the ewe in relation to LH secretion. *Neuroendocrinology* **39**, 256–260.

Wilson, R.C., Kesner, J.S., Kaufman, J.M., Uemura, T., Akema, T. & Knobil, E. (1984). Central electrophysiological correlates of pulsatile luteinizing hormone secretion in the rhesus monkey. *Neuroendocrinology* **39**, 256–260.

Yeates, T.M. (1949). The breeding season of the sheep with particular reference to its modification by artificial means using light. *Journal of Agricultural Science* **39**, 1–43.

Yoshioka, Z. (1961). Studies on the artificial control of the sexual activity in ewes and she-goats. *Japanese Journal of Animal Reproduction* **7**, 93–102.

6 Neuroendocrine Mechanism Regulating the Pulsatile Luteinizing Hormone Secretion

Kei-ichiro MAEDA, Hiroko TSUKAMURA, Satoshi OHKURA and Akira YOKOYAMA

School of Agricultural Sciences, Nagoya University, Nagoya 464-01

It has been widely accepted that there are two centers for luteinizing hormone (LH) release in the brain (Gorski, 1968): one is the tonic center located in the mediobasal hypothalamus (MBH) regulating tonic LH secretion which is responsible for follicular development and the other is the cyclic center located in the medial preoptic area (POA) regulating LH surge which induces ovulation.

Over the last decades, developments of the techniques for frequent blood samplings and sensitive assays for LH have enabled us to reveal that both types of LH release consist of pulses: The tonic basal LH secretion shows pulsatile fluctuations at regular intervals (Gallo, 1981b) and even the LH surge is composed of many pulses (Gallo, 1981a). Since each LH pulse corresponds to a LH-releasing hormone (LHRH) pulse which is observed in blood of the portal vessels (Levine *et al.*, 1982; Levine & Duffy, 1988), pulsatile LH release would be regulated by pulsatile LHRH release. The brain mechanism generating LH pulses has, therefore, been called the "LHRH pulse generator" (Lincoln *et al.*, 1985; Dyer & Robinson, 1989).

It has been demonstrated that frequency is the most important component of LH pulses in controlling the activity of the gonadal axis. For instance, environmental factors have been shown to suppress or activate the gonadal function by altering the frequency of pulses of tonic LH secretion. Long days reduce and short days increase the frequency of LH pulses in rams and ewes (Lincoln & Short, 1980; Karsch *et al.*, 1984), and thus photoperiodic cues alter the gonadal activity. Physical stresses also dramatically reduce the pulsatility of LH secretion (Higuchi *et al.*, 1986; Briski & Sylvester, 1988). In rats, the frequency of LH pulses is profoundly reduced by 48-h food deprivation in an estrogen-dependent manner (Cagampang *et al.*, 1990; Cagampang *et al.*, 1991) and by the suckling stimulus without ovarian steroids (Maeda *et al.*, 1989). On the other hand, the activity of the mechanism for LH surges does not seem to be affected by the environmental factors, since a LH surge can be induced by the administration of estradiol (E_2) even in seasonally anestrous (Mori *et al.*, 1987) or lactating (Tsukamura *et al.*, 1988) animals. These suggest that most environmental factors impact on the gonadal activity by regulating the pulsatile basal LH secretion, especially by modulating its frequency, but not the surge-like LH secretion.

The location of the center for tonic LH release has been intensively investigated. Halasz & Pupp (1965) first demonstrated in rats that persistent estrus appeared after complete denervation of afferent fibers into MBH with a bayonet-shaped knife and concluded that the tonic center for LH release is located in the island made by the complete deafferentation. This was confirmed by many researchers using a Halasz's type knife, including Blake & Sawyer (1974) who showed that pulsatile LH release was still apparent in rats bearing the complete deafferentation. However, the development of immunocytochemistry of LHRH has revealed that cell bodies of LHRH-producing neurons are distributed in the area anterior to the MBH but not in the MBH (Kawano & Daikoku, 1981; Witkin *et al.*, 1987). Some researchers, therefore, have been suspicious of the location of the tonic center for LH release in the MBH. They believe that the mechanism generating LHRH pulses is outside of the hypothalamus and that monoaminergic or other neural input to LHRH perikarya, located mainly in POA, is generating LHRH pulses (Coen, 1987; Wuttke *et al.*, 1987). Others think that the LHRH neuron itself is equipped with an intrinsic pulse generator.

One point of controversy may be due to the unclear definition of the term, "LHRH pulse generator". To further the discussion, we postulate that the LHRH pulse generator is a group of neurons which controls an intermittent release of LHRH from the nerve endings of LHRH neurons into the portal vessels. We first describe the importance of the pulsatile LH secretion for the environmental factors to regulate the activity of gonadal function, based on results obtained in the lactating animal model. We then introduce our recent results and discuss the possible location and mechanism of the LHRH pulse generator.

THE SUCKLING-INDUCED SUPPRESSION OF THE PULSATILE LH SECRETION

Ovarian cyclicity ceases during lactation in many mammalian species including human (Smith *et al.*, 1990; Maeda *et al.*, 1991; for reviews see McNeilly, 1988). The cessation of this cyclicity may be primarily caused by the suckling of teats by pups or babies, the suckling stimulus, since the decrease in frequency and intensity of the suckling stimulus results in an earlier recurrence of the normal estrous or menstrual cycles (Howie & McNeilly, 1982; Howie *et al.*, 1982; Stevenson & Britt, 1981; Stevenson *et al.*, 1981; Taya & Sasamoto, 1980).

How is the Pulsatile LH Secretion Suppressed by the Suckling Stimulus?
In rats, ovulation is usually blocked for the first 20 days after the post-partum ovulation which occurs about 30 hours after parturition (Tomogane *et al.*, 1976). High levels of circulating progesterone (Tomogane *et al.*, 1969), which is secreted by the lactational corpora lutea formed after the post-partum ovulation, have been believed to play a crucial role in suppressing the LH secretion in lactating rats. We first attempted to demonstrate whether the pulsatile LH secretion would be suppressed by the suckling stimulus without the presence of ovarian steroids in lactating rats. We used lactating rats ovariectomized on day 2 of lactation (the day of parturition was designated day 0 of lactation). Lactation proceeds normally in ovariectomized rats.

The pulsatile LH secretion was profoundly suppressed at mid-lactation in ovariectomized lactating rats (Maeda *et al.*, 1987). LH pulses were apparent in the late lactation, suggesting that the suppressive effect of the suckling stimulus becomes weaker with the advancement of lactation. When pups were separated from the mother at early lactation, the 3 parameters, the mean LH concentration over the 3-h sampling period and the frequency and amplitude of LH pulses, began to increase from 12 h after the separation of pups (Figs. 1 and 2). Of these 3 parameters, the frequency of LH pulses, which might directly reflect the release of LHRH, dramatically increases after the separation of pups (Fig. 2). On the contrary, the frequency of LH pulses did not change after ovariectomy in cycling rats (Fig. 3). These results suggest that the suckling stimulus suppresses the LH secretion in a different manner from that of ovarian steroids.

The suckling stimulus did not affect the incidence of daily LH surges

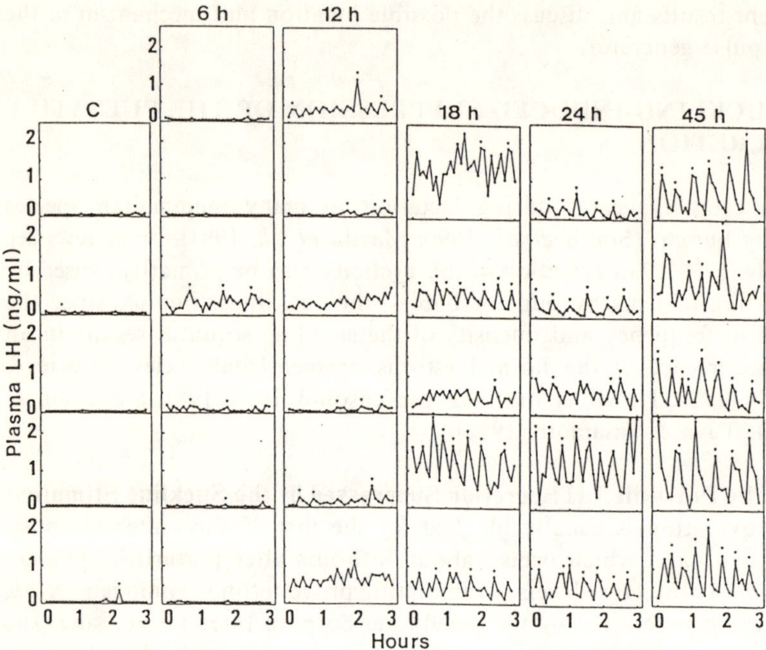

Fig. 1. Plasma profiles of LH in individual ovariectomized lactating rats deprived of their pups for 6, 12, 18, 24 or 45 h. Litter size was adjusted to eight on day 1 of lactation. All rats were ovariectomized on day 2 of lactation and blood samples were taken on day 8 of lactation every 6 min for 3 h (13.00–16.00 h). The values were expressed in terms of the NIDDK-rLH-RP-2. Arrowheads indicate the LH pulses identified with the PULSAR computer program (Merriam & Wachter, 1982). C: control lactating rats which had not been deprived of their pups. (From Maeda *et al.*, 1989).

by chronic E_2 treatments (Fig. 4). These surges occur in E_2-implanted ovariectomized lactating rats as well as in non-lactating controls, although the lactating group shows a steeper decline in the amplitude of daily LH surges. This suggests that the suckling stimulus does not impair the mechanism inducing LH surges.

Most of the environmental factors, such as photoperiod, fasting and stress, influence the LH secretion by altering the threshold of sensitivity of the LH-releasing mechanism to the negative feedback effects of estrogen (Karsch *et al.*, 1980; Cagampang *et al.*, 1991; Briski & Sylvester, 1988). Our studies show that the suckling stimulus inhibits the pulsatile LH secretion, particularly the frequency of LH pulses, in the absence of ovarian steroids. In addition, the suckling stimulus does not affect the positive feedback mechanism of estrogen that is responsible for inducing LH surges. These

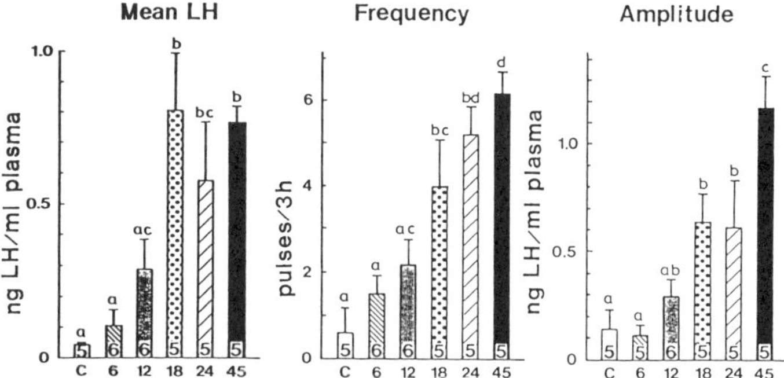

Fig. 2.   The mean LH level and the frequency and amplitude of LH pulses calculated by the PULSAR computer program in ovariectomized lactating rats deprived of their pups for 6, 12, 18, 24 or 45 h. Values are means±SEM. Mean LH levels represent the means for all samples collected over the 3-h sampling period. Numbers in each column represent the number of animals used. Values with different letters are significantly different from each other ($P<0.01$, Duncan's multiple-range test). C: control lactating rats which had not been deprived of their pups. See Fig. 1 for further details. (From Maeda *et al.*, 1989).

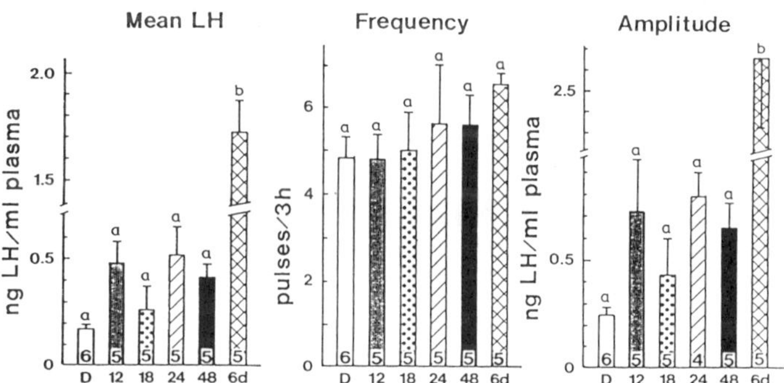

Fig. 3.   The mean LH level and the frequency and amplitude of LH pulses in rats ovariectomized for 12, 18, 24, 48 h or 6 days. Cycling female rats were ovariectomized at diestrus. D: control rats bled on day 1 of diestrus. See Figs. 1 and 2 for further details. (From Maeda *et al.*, 1989).

findings suggest that the suckling stimulus is directly involved in suppressing the activity of the mechanism generating LH pulses, the LHRH pulse generator in a steroid-independent manner.

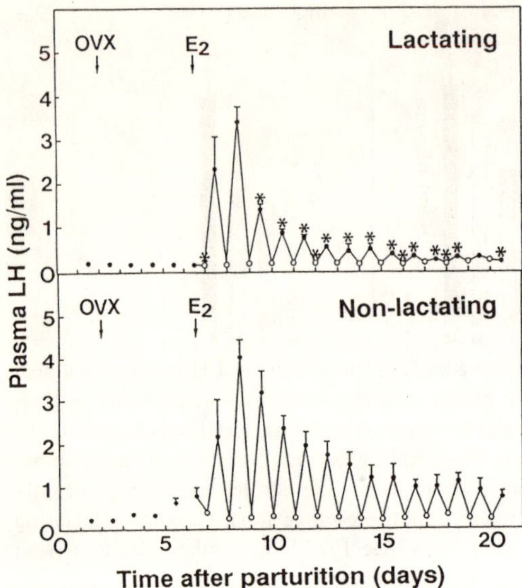

Fig. 4. Daily LH surges in ovariectomized (a) lactating and (b) non-lactating rats ($n =$ 8 in each group) implanted with silicone tubing containing crystalline E_2 on day 6. Litter size was adjusted to eight on day 1 in lactating rats. In non-lactating rats, pups were removed from their mothers on day 0. Ovariectomy was performed on day 2 post-partum. Open and closed circles indicate the values at 10.00 and 17.00 h, respectively. $*P < 0.05$ compared with non-lactating rats (Student's t-test). Values are means $\pm$ SEM. Arrows indicate the times of ovariectomy (OVX) and E_2 implantation (E_2). (From Tsukamura *et al.*, 1988).

Neural Pathways Mediating the Suppressive Effect of the Suckling Stimulus on the Activity of the LHRH Pulse Generator

Our previous studies showed that some inhibitory neural pathways seem to be directly involved in suppressing the activity of the LHRH pulse generator in the hypothalamus. We, therefore, examined the effect of various hypothalamic deafferentations on the suppressed secretion of LH in ovariectomized lactating rats to reveal pathways of the signals emanating from teats suckled by pups and inhibiting the activity of the LHRH pulse generator in the hypothalamus.

Hypothalamic deafferentations employed in our experiments were complete (CD), anterior (AD), anterolateral (ALD), posterior (PD), roof (RD) and sham- (SD) deafferentations shown in Fig. 5. Figure 6 and Table 1 show, respectively, the representative profiles and the parameters of LH pulses in ovariectomized lactating rats bearing various deafferentations at early lactation. CD, ALD and RD significantly increased the parameters of

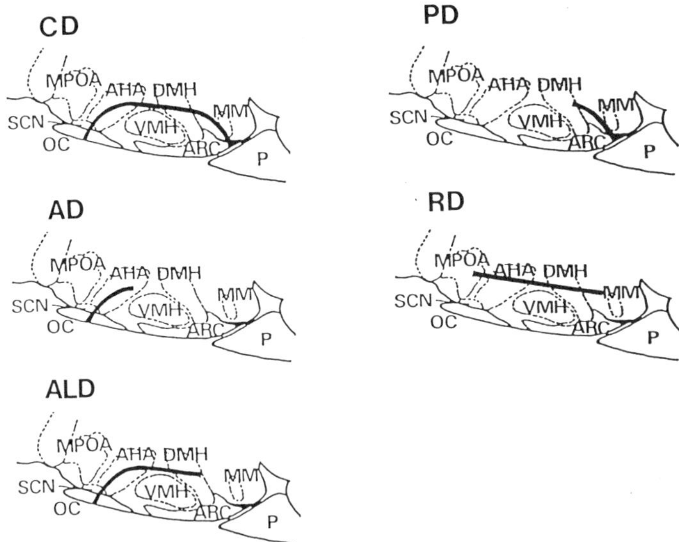

Fig. 5. Schematic illustrations of the sites of hypothalamic deafferentations in a parasagittal plane. CD, complete; AD, anterior; ALD, anterolateral; PD, posterior; RD, roof deafferentation; AHA, anterior hypothalamic area; ARC, arcuate nucleus; DMH, dorsomedial hypothalamic nucleus; MM, mammillary body; MPOA, medial preoptic area; OC, optic chiasm; P, anterior pituitary; SCN, suprachiasmatic nucleus; VMH, ventromedial hypothalamic nucleus. (From Tsukamura *et al.*, 1990).

TABLE 1

Mean plasma LH and prolactin (PRL) concentrations and frequency and amplitude of LH pulses (means±SEM.) in ovariectomized lactating rats bearing various hypothalamic deafferentations

	n[g]	Mean LH (ng/ml)	LH pulse frequency (pulses/3 h)	LH pulse amplitude[h] (ng/ml)		Mean PRL (ng/ml)
SD[a]	5	0.07±0.00	1.40±0.40	0.14±0.02	(4)[i]	63.8±17.5
CD[b]	7	0.73±0.17**	5.71±0.81*	0.52±0.08**	(7)	74.4±11.4
AD[c]	4	0.24±0.19	2.25±2.25	0.50	(1)	62.2±31.7
ALD[d]	5	0.78±0.27**	4.00±1.38	0.48±0.07*	(4)	67.7±29.7
PD[e]	7	0.04±0.00**	0*	—	(0)	64.4±18.2
RD[f]	5	0.66±0.13**	8.00±0.63	0.44±0.07	(5)	20.1±1.0**

[a] Sham-, [b] complete, [c] anterior, [d] anterolateral, [e] posterior and [f] roof deafferentation. [g] Number of animals used. [h] Pulse amplitudes were calculated in animals showing LH pulses. [i] Number of animals showing LH pulses.
*$P<0.05$, **$P<0.01$ compared with SD (Mann-Whitney U-test). (From Tsukamura *et al.*, 1990).

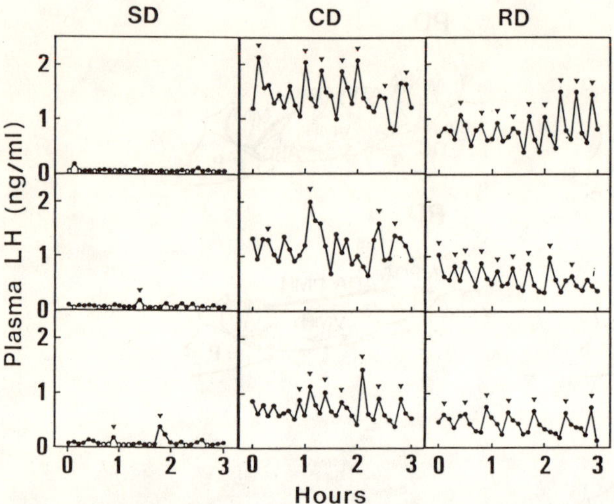

Fig. 6.   Representative profiles of plasma LH concentrations in individual ovariectomized lactating rats bearing sham- (SD), complete (CD) and roof (RD) deafferentations on day 7 or 8 of lactation. Litter size was adjusted to eight on day 1 and ovariectomy was performed on day 2 of lactation. Blood sampling was started 24 h after hypothalamic deafferentation performed on day 6 or 7 of lactation. Arrowheads represent the peaks of LH pulses identified by the PULSAR computer program. (From Tsukamura *et al.*, 1990).

LH pulses compared to SD (Table 1): LH pulses were apparent in all individuals (Fig. 6). Nevertheless, AD and PD did not affect the pulsatile LH secretion: All the parameters of LH pulses were kept at a very low level as observed in mothers with SD (Table 1).

These results clearly indicate that the inhibitory signal emanating from the suckling stimulus is conveyed dorsally to the hypothalamus, since only the deafferentations which cut the dorsal part of the hypothalamus (CD, ALD and RD) were able to block the suppressive effect of the suckling stimulus on LH pulses. These findings also imply that the signal inhibits the activity of the LHRH pulse generator possibly located in the MBH, since LH pulses were still apparent in animals bearing CD which isolated the MBH from the rest of the brain. Plasma prolactin levels were not affected by any deafferentations except for RD by which plasma prolactin levels significantly decreased (Table 1), suggesting that the suppression of LH pulses is not necessarily associated with prolactin secretion (Maeda *et al.*, 1990; Tsukamura *et al.*, 1991), and that prolactin secretion might be autonomously controlled within the MBH in the absence of the afferent input to the hypothalamus during lactation.

Our preliminary study on the more detailed pathways using various sizes and depth of RD and electrolytic lesion of the paraventricular nucleus (PVN) showed that RD partly blocked the suppressive effect of the suckling stimulus only when it cut the ventral margin of the PVN. In addition, the electrolytic lesion of the PVN partly restored the suppressed pulsatile LH secretion only when the lesion was extended to the periventricular region, indicating that the PVN is apparently not involved in the neural pathway and that the periventricular nucleus plays a role in mediating the inhibitory effect of the suckling stimulus on pulsatile LH secretion.

In conclusion, the information which emanates from the suckling stimulus and inhibits LH pulses is conveyed dorsally to the MBH *via* neural fibers possibly passing through the periventricular region but not through the PVN. It should be emphasized that a mechanism generating LH pulses could be located in the MBH, because LH pulses became apparent in ovariectomized lactating rats with complete deafferentation which denervated all the neural pathways entering the MBH.

THE LOCATION OF THE LHRH PULSE GENERATOR IN THE HYPOTHALAMUS

Our studies on the hypothalamic deafferentation in ovariectomized lactating rats support Halasz & Pupp's hypothesis (1965) that the mechanism regulating the tonic LH release responsible for follicular development is located in the MBH. Blake & Sawyer (1974) demonstrated that pulsatile LH secretion was still apparent after complete deafferentation. In their studies, however, some animals showed persistent diestrus and no apparent LH pulses. Why did the response to hypothalamic deafferentation vary from animal to animal in their previous studies? Is 'complete' deafferentation sometimes incomplete?

We have done several experiments to answer these questions using chronically ovariectomized rats to exclude the influence of ovarian steroids.

Effect of Various Hypothalamic Deafferentations on the Pulsatile LH Release in Ovariectomized Rats

We first examined the effect of various types of hypothalamic deafferentations, AD, ALD and CD (Fig. 5) on the pulsatile LH secretion in ovariectomized rats (Figs. 7 and 8). No type of deafferentation affects the frequency of LH pulses (Fig. 8), which might directly reflect the activity of the LHRH pulse generator, whereas the amplitudes of LH pulses, which seem to reflect the amount of LHRH released into the portal vessel, are decreased with the posterolateral extension of the incision of the deafferen-

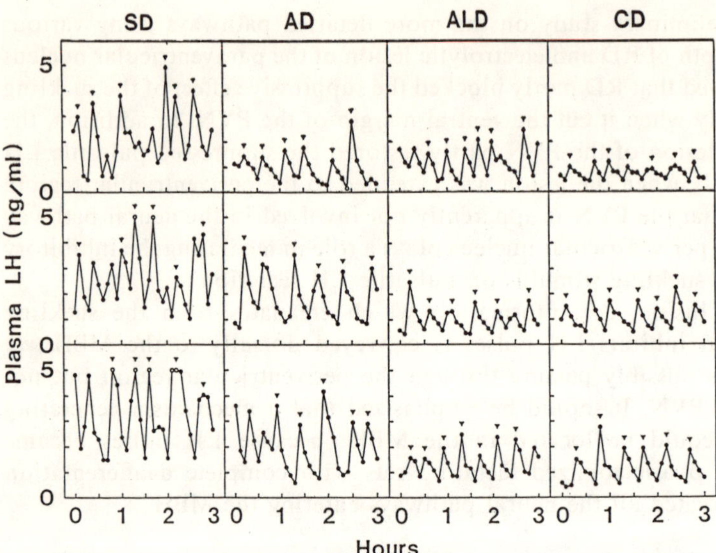

Fig. 7. Pulsatile LH secretion in 3 representative ovariectomized rats of each group bearing SD, AD, ALD and CD. Rats ovariectomized for 1 week were subjected to each hypothalamic deafferentation and blood samples were collected 5 days after the surgery at 6-min intervals for 3 h. Arrowheads represent the LH pulse identified by the PULSAR computer program. (From Ohkura *et al.*, 1991).

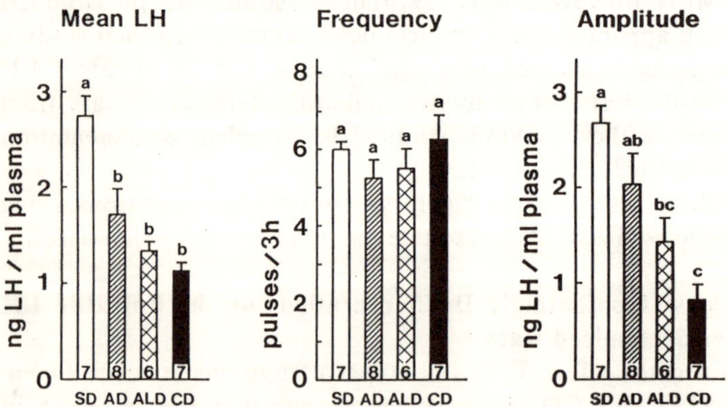

Fig. 8. Parameters of the pulsatile LH secretion in rats bearing SD, AD, ALD and CD. Mean LH levels were calculated for a series of plasma LH concentrations during the 3-h sampling period. The frequency and amplitude of LH pulses were identified by the PULSAR computer program. Values are means±SEM. Numbers in each column indicate the number of animals used. Values with different letters are significantly different from each other ($P < 0.05$, Duncan's multiple-range test). (From Ohkura *et al.*, 1991).

tation. These results indicate that the LHRH pulse generator could be located in the hypothalamic island made by CD, namely MBH, and thus its activity could not be affected by any type of deafferentation. On the other hand, it is likely that the more posterolaterally the incision extends, the more LHRH fibers are cut. This would result in a lesser amount of LHRH being released into the portal vessels and, consequently, lower amplitude of LH pulses would appear. These findings support the hypothesis proposed by Halasz & Pupp (1965) that the center for the tonic LH secretion is located in the MBH.

Verification of the Location of the LHRH Pulse Generator Using the Transplantation of Brain Tissues

There is still a debate over whether the pulsatile LH secretion in animals

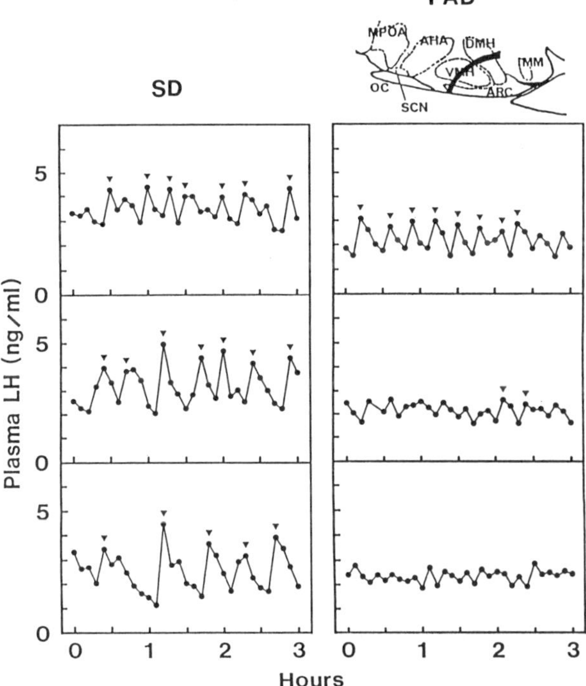

Fig. 9. Representative profiles of plasma LH concentrations in ovariectomized rats bearing SD and posterior-anterior deafferentation (PAD). Chronically (>2 weeks) ovariectomized rats received PAD 1 week before the blood collection. Arrowheads indicate LH pulses identified with the PULSAR computer program. Schematic illustration of a sagittal section of the hypothalamus with the position of PAD is also shown (top right). (From Ohkura *et al.*, 1991).

bearing CD is actually due to the incompleteness of CD (this expression is somewhat paradoxical). Some researchers believe that CD does not cut all the LHRH fibers passing through the optic chiasm, *i.e.* the subchiasmatic LHRH fibers (Coen, 1987; a few LHRH fibers usually observed under the optic chiasm) are spared, and thus the LHRH pulse generator located outside of the MBH, which might be the LHRH neuron itself or some neurons innervated to the LHRH perikarya, is able to generate pulses after CD. To verify the location of the LHRH pulse generator in the MBH, we employed posterior-anterior deafferentation (PAD), which severs the anterior part of the arcuate nucleus (ARC) from the MBH and was first introduced by Blake & Sawyer (1974).

LH pulses became much less apparent after PAD compared to those in sham-deafferentated animals (Fig. 9): Fluctuations of plasma LH levels were irregular and small and did not seem to be the normal pulses. On the other hand, the baseline level of LH pulses was not suppressed completely, indicating that LHRH release might not be suppressed completely after PAD. Our immunohistochemical screening revealed that a considerable number of LHRH fibers was spared by the PAD (Fig. 10). These results suggest that the mechanism involved in the induction of LHRH pulses, possibly the LHRH pulse generator, is located in the anterior part of the ARC. The mechanism does not seem to include LHRH neurons themselves, because if LHRH neurons had an intrinsic pulse-generating mechanism, LHRH pulses and hence LH pulses should occur in animals bearing PAD which spared many LHRH neuronal fibers.

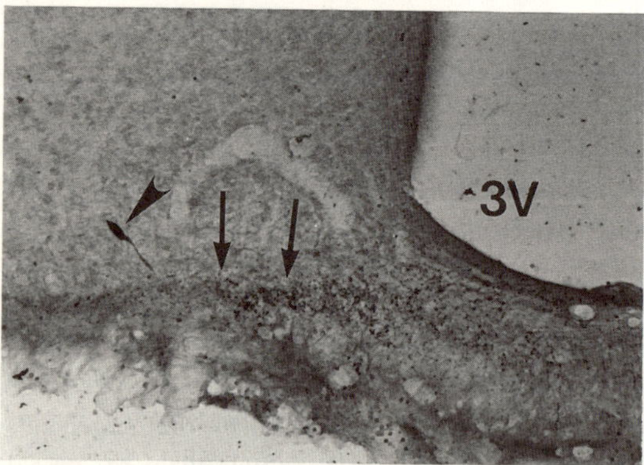

Fig. 10.   A photomicrograph showing LHRH-positive neuronal fibers (arrows) in the median eminence of an ovariectomized rat with PAD. An arrowhead indicates the LHRH-positive neuronal cell body adjacent to the lateral border of the median eminence. 3V, third ventricle. Magnification: ×112.5. (From Ohkura *et al.*, 1991).

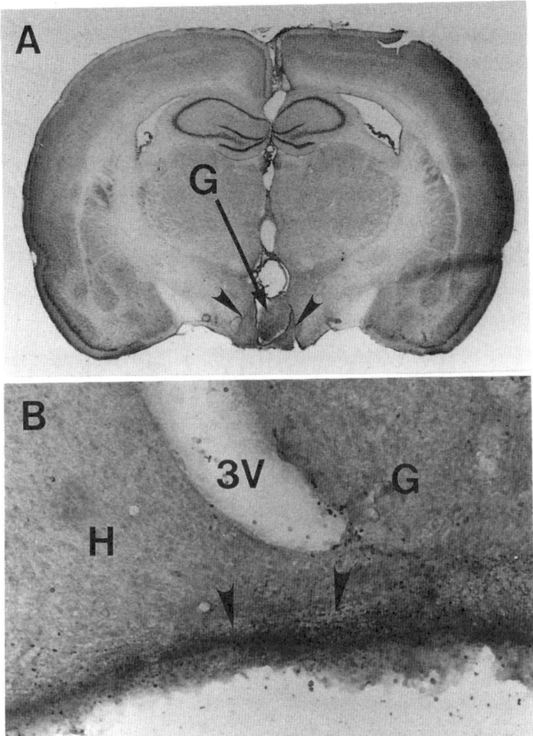

Fig. 11. Photomicrographs of frontal sections from the brain of an ovariectomized rat which had received a transplant of fetal MBH tissue and PAD. (A) MBH graft in the third ventricle. Arrowheads indicate the location of the knife cuts. (B) LHRH-positive fibers (arrowheads) in the median eminence. No LHRH-positive neuronal cell body was found in the MBH graft. G: graft, H: host, 3V: third ventricle. Magnifications: (A) ×4.5, (B) ×125. (From Ohkura *et al.*, 1992).

We verified the above hypothesis by transplanting the MBH tissue into the 3rd ventricle of animals followed by PAD. We transplanted 17-day-old fetal brain tissue into the 3rd ventricle of ovariectomized rats 4 weeks prior to PAD to ensure that the transplanted tissue survived in the host brain (Fig. 11A). Blood samples were taken 1 week after PAD, since we had found in preliminary experiments that normal pulses were observed 5 weeks but never 1 week after PAD (Ohkura *et al.*, 1991).

The fetal MBH tissue transplanted prior to PAD maintained normal LH pulses which were impaired by PAD in non-transplanted animals or cortex-transplanted controls (Figs. 12 and 13). No LHRH immuno-reactivity was found in fetal MBH tissues surviving in the 3rd ventricle of the

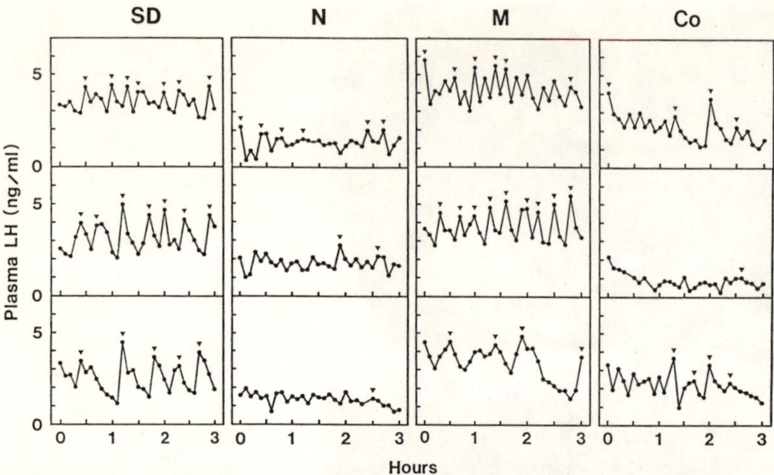

Fig. 12.   Profiles of plasma LH concentrations in 3 representative ovariectomized rats following SD or PAD (N, M and Co). Four weeks prior to PAD, ovariectomized (>2 weeks) rats received either no transplant (N), fetal MBH (M) or fetal cerebral cortex (Co) transplants in the 3 rd ventricle. Blood was collected 1 week after PAD. Arrowheads represent the LH pulse identified by the PULSAR computer program. (From Ohkura *et al.*, 1992).

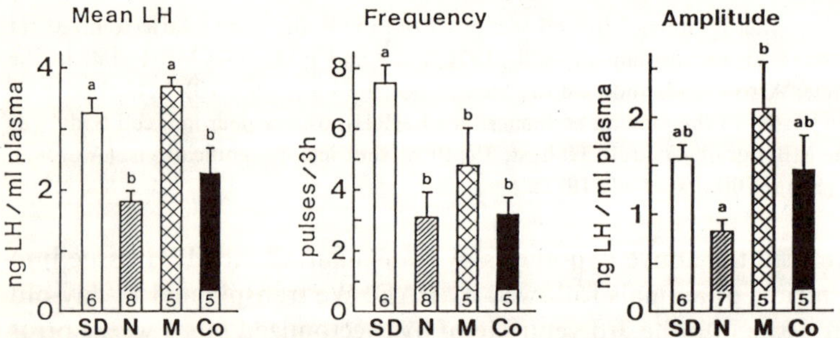

Fig. 13.   LH pulse characteristics in ovariectomized rats following SD or PAD (N, M and Co). Prior to PAD rats received either no transplant (N), fetal MBH (M) or fetal cerebral cortex (Co) transplants. Mean LH levels were calculated for a series of plasma LH concentrations during the 3-h sampling period. The frequency and amplitude of LH pulses were identified by the PULSAR computer program. Values are means±SEM. Numbers in each column indicate the number of animals used. The amplitude of LH pulses was calculated only in rats in which LH pulses were identified by the PULSAR computer program. Values with different letters are significantly different from each other (*P*<0.05, Duncan's multiple-range test). (From Ohkura *et al.*, 1992).

host brain (Fig. 11B). These results demonstrated that some mechanism, other than LHRH neurons in the fetal MBH is crucial in maintaining regular LH pulses. We, therefore, conclude that the LHRH pulse generator consists of non-LHRH neurons and could be located in the MBH, particularly in the anterior part of the ARC.

CONTROVERSY OVER THE LOCATION OF THE LHRH PULSE GENERATOR

There have been several bodies of evidence supporting our idea that the LHRH pulse generator is located in the ARC. A periodic rise in the multiple unit activity (MUA), which is in synchrony with the peak of LH pulses, is recorded in the anterior portion of the ARC but never in the POA in ovariectomized rats (Kawakami *et al.*, 1982; see Chapters 4 and 5). In monkeys in which most of LHRH perikarya are located in the MBH (Silverman *et al.*, 1979), a similar elevation of MUA is observed in the area like that in rats (Wilson *et al.*, 1984). It is likely that the MUA rise in the MBH reflects the excitation of a group of neuronal cell bodies in this area. However, another possibility can not be dismissed that the rise is due to the excitation of axons of LHRH neurons concentrated in this region in both rats and monkeys regardless of the distribution of LHRH neuronal cell bodies. Soper & Weick (1980) reported that an electrolytic lesion of the ARC combined with AD abolishes LH pulses in ovariectomized rats but a lesion of the ARC alone does not. They concluded that at least one pulse generator would be located in the ARC and the others would be extra-hypothalamic.

Some researchers, in contrast, consider that the LHRH pulse generator is located in the POA or outside of the hypothalamus and controls LHRH release at the POA (see Dyer & Robinson, 1989). There has been reported an inverse relationship between the pulsatile release of γ-aminobutyric acid (GABA) in the POA and LH secretion (Wuttke *et al.*, 1987). In addition, the local application of a specific α_1-adrenergic agonist into POA but not into MBH reduces the LH pulsatility (Jarry *et al.*, 1990). These results suggest that the adrenergic input from the brain stem to POA or GABAergic mechanisms inside the POA is involved in regulating LH pulses. Our results, on the contrary, showed that PAD, which spared most of the LHRH neuronal fibers, impaired LH pulses, suggesting that LHRH neurons do not themselves have an intrinsic pulse-generating mechanism. It is most likely that PAD cuts the pathway from possible LHRH pulse-generating mechanisms, which might be located in the ARC, to LHRH-producing neurons.

POSSIBLE MECHANISM REGULATING PULSATILE LHRH RELEASE

LHRH is apparently released simultaneously from all the LHRH-producing neuronal terminals into the portal vessels to form a clear peak at a given interval. How is the LHRH pulse generated in the hypothalamus? There are two possible regulation sites of pulsatile LHRH release: One is at the LHRH neuronal cell bodies and the other is at their terminals. A dense network is anticipated either between LHRH cell bodies themselves or between their cell bodies and other neuronal terminals, if pulsatile LHRH release is regulated at the cell bodies. In oxytocinergic neurons, glial processes between oxytocin neuronal cell bodies disappear and they are densely connected with each other by double synapses during lactation (Theodosis *et al.*, 1986) when oxytocin is released with a clear pulsatility (Chapter 8). Although there have been reported synaptic contacts of several neuronal terminals with the LHRH perikarya in the POA (Leranth *et al.*, 1985; Pelletier, 1987; Leranth *et al.*, 1988; MacLusky *et al.*, 1988), such a contact is not often observed. In addition, LHRH cell bodies are distributed too diffusely over a wide area of the hypothalamus to be densely connected with other neurons (Kawano and Daikoku, 1981; Witkin *et al.*,

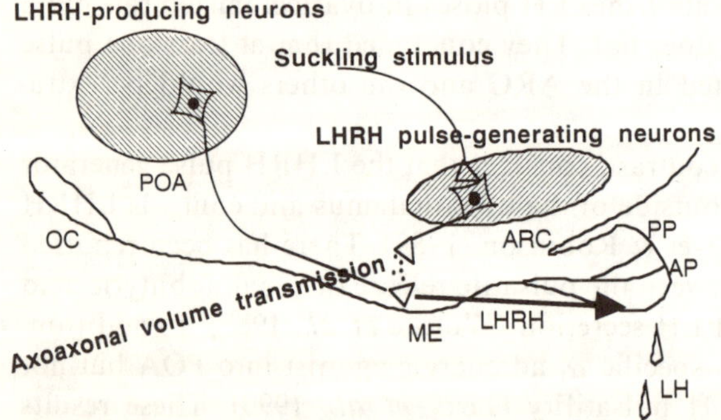

Fig. 14.   Possible mechanism of generation of the pulsatile LHRH release in the brain. Cell bodies of LHRH-producing neurons are located extensively in the preoptic/septum region (POA) with their axon projecting to the median eminence (ME). The axonal terminals of LHRH pulse-generating neurons, whose perikarya are located in the arcuate nucleus (ARC), contact the terminals of LHRH-producing neurons non-synaptically and regulate the pulsatile LHRH release at the ME (axoaxonal volume transmission). The suckling stimulus enters the hypothalamus dorsally and inhibits the activity of LHRH pulse-generating neurons. AP, anterior pituitary; OC, optic chiasm; PP, posterior pituitary.

1982) and assure the synchronized release of LHRH. Therefore, it is more likely that LHRH release is controlled at the neuronal terminals in the median eminence rather than at the cell bodies (Fig. 14). Since LHRH terminals are converged and concentrated in the median eminence, neural inputs from the pulse generator can easily contact all the terminals and synchronize LHRH release. Few synapses have been found in the median eminence (Kobayashi & Matsui, 1969) in spite of the abundance of receptors to various neurotransmitters in this area (Beauvillain *et al.*, 1991), allowing us to postulate that LHRH release is regulated non-synaptically in the median eminence. Recently, a novel neurotransmission mechanism, called "volume transmission", was proposed by Fuxe (1991) in opposition to "wiring transmission". Leakage of neurotransmitters has often been found at a site where no synapse is observed, indicating that "synapse" is not the only means of communication between neurons. It seems that volume transmission is more useful under a certain condition where neurons control the activity of many adjacent neurons simultaneously at a slow interval (20-30 min) as is seen in LHRH release.

PERSPECTIVES

Most pituitary hormones, *i.e.* follicle-stimulating hormone (FSH), growth hormone (GH), and prolactin are secreted episodically, suggesting that the hypothalamic mechanisms regulating the secretion of these hormones are activated periodically. Is LHRH the only hypothalamic hormone which has a pulse generator? Do other hypothalamic hormones have their own pulse generators? LHRH-producing cell lines have recently been established from the brain of transgenic mice that were infected with LHRH genes (Weiner *et al.*, 1990). These cell lines secrete LHRH into the medium *in vitro* in a pulsatile manner though the frequency is much higher than that *in vivo*. The cell lines include neurons which would not express the LHRH gene in normal mice. However, the pulsatile LHRH release from these cells suggests that each cell has its own secretory oscillation. It is likely that the oscillation of each neuron is integrated in a different way from hormone to hormone, allowing hormones to be secreted at different intervals. The mechanism governing the periodic release of those hormones might, therefore, imply that there is a fundamental mechanism integrating the oscillations in individual cells.

CONCLUSION

The suckling stimulus is a novel stimulus inhibiting LH pulses without any cooperation from ovarian steroids, unlike other stimuli (stress, photoperiod, fasting, *etc.*). It is directly involved in suppressing the activity of the LHRH pulse generator, since the frequency of LH pulses is suppressed by the suckling stimulus. The information emanating from the teats suckled by pups or babies and inhibiting the pulsatile LH secretion is conveyed dorsally to the MBH, where the LHRH pulse generator is located. A part of these inhibitory fibers pass through the periventricular region but not the PVN. Experiments using various deafferentations and transplantation of fetal brain tissue revealed that the LHRH pulse generator is located in the anterior part of the ARC. It probably consists of a group of non-LHRH-producing neurons and regulates the pulsatile LHRH release at the terminals of LHRH-producing neurons in the median eminence non-synaptically.

REFERENCES

Beauvillain, J.C., Moyse, E., Dutriez, I., Mitchell, V., Poulain, P. & Mazzuca, M. (1991). Visualization of membrane-associated mu-opioid receptors in the guinea pig median eminence. *Abstracts of the Third IBRO World Congress of Neuroscience*, p. 383.

Blake, C.A. & Sawyer, C.H. (1974). Effects of hypothalamic deafferentation on the pulsatile rhythm in plasma concentrations of luteinizing hormone in ovariectomized rats. *Endocrinology* **94**, 730-736.

Briski, K.P. & Sylvester, P.W. (1988). Effect of specific acute stressors on LH release in OVX and OVX-E_2 treated female rats. *Neuroendocrinology* **47**, 194-202.

Cagampang, F.R.A., Maeda, K.-I., Tsukamura, H., Ohkura, S. & Ôta, K. (1991). Involvement of ovarian steroids and endogenous opioids in the fasting-induced suppression of pulsatile LH release in ovariectomized rats. *Journal of Endocrinology* **129**, 321-328.

Cagampang, F.R.A., Maeda, K.-I., Yokoyama, A. & Ôta, K. (1990). Effect of food deprivation on the pulsatile LH release in the cycling and ovariectomized female rat. *Hormone and Metabolic Research* **22**, 269-272.

Coen, C.W. (1987). Hypothalamic biogenic amines and the regulation of luteinizing hormone release in the rat. *Endocrinology and Physiology of Reproduction* (Leung, P.C.K., Armstrong, D.T., Ruf, K.B., Moger, W.H. and Friesen, H.G., eds.), Plenum Press, New York, pp. 71-98.

Dyer, R.G. & Robinson, J.E. (1989). The LHRH pulse generator. *Journal of Endocrinology* **123**, 1-2.

Fuxe, K. & Agnati, L.F. (1991). *Volume Transmission in the Brain,* Raven Press, New York.

Gallo, R.V. (1981a). Pulsatile LH release during the ovulatory LH surge on proestrus in the rat. *Biology of Reproduction* **24**, 100-104.

Gallo, R.V. (1981b). Pulsatile LH release during periods of low level LH secretion in the rat estrous cycle. *Biology of Reproduction* **24**, 771-777.

Gorski, R.A. (1968). The neural control of ovulation. *Biology of Gestation* (Assali, N.S., ed.), Academic Press, New York. pp. 1-66.

Hálasz, B.&.Pupp, L. (1965). Hormone secretion of the anterior pituitary gland after physical interruption of all nervous pathways to the hypophysiotropic area. *Endocrinology* **77**, 553–562.

Higuchi, T., Honda, K. & Negoro, H. (1986). Influence of oestrogen and noradrenergic afferent neurons on the response of LH and oxytocin to immobilization stress. *Journal of Endocrinology* **110**, 245–250.

Howie, P.W. & McNeilly, A.S. (1982). Effect of breast-feeding patterns on human birth intervals. *Journal of Reproduction and Fertility* **65**, 545–557.

Howie, P.W., McNeilly, A.S., Houston, M.J., Cook, A. & Boyle, H. (1982). Fertility after childbirth: Post-partum ovulation and menstruation in bottle and breast feeding mothers. *Clinical Endocrinology* **17**, 323–332.

Jarry, H., Leonhardt, S. & Wuttke, W. (1990). A norepinephrine-dependent mechanism in the preoptic/anterior hypothalamic area but not in the mediobasal hypothalamus is involved in the regulation of the gonadotropin-releasing hormone pulse generator in ovariectomized rats. *Neuroendocrinology* **51**, 337–344.

Karsch, F.J., Bittman, E.L., Foster, D.L., Goodman, R.L., Leagan, S.J. & Robinson, J.E. (1984). Neuroendocrine basis of seasonal reproduction. *Recent Progress in Hormone Research* **40**, 185–232.

Karsch, F.J., Goodman, R.L. & Leagan, S.J. (1980). Feedback basis of seasonal breeding: Test of an hypothesis. *Journal of Reproduction and Fertility* **58**, 521–535.

Kawakami, M., Uemura, T. & Hayashi, R. (1982). Electrophysiological correlates of pulsatile gonadotropin release in rats. *Neuroendocrinology* **35**, 63–67.

Kawano, H. & Daikoku, S. (1981). Immunohistochemical demonstration of LHRH neurons and their pathways in the rat hypothalamus. *Neuroendocrinology* **32**, 179–186.

Kobayashi, H. & Matsui, T. (1969). Fine structure of the median eminence and its functional significance. *Frontiers in Neuroendocrinology*, (Ganong, W.F. and Martini, L., eds.), Oxford University Press, New York, pp. 3–46.

Leranth, C., MacLusky, N.J., Sakamoto, H., Shanabrough, M. & Naftolin, F. (1985). Glutamic acid decarboxylase-containing axons synapse on LHRH neurons in the rat medial preoptic area. *Neuroendocrinology* **40**, 536–539.

Leranth, C., MacLusky, N.J., Shanabrough, M. & Naftolin, F. (1988). Catecholaminergic innervation of luteinizing hormone-releasing hormone and glutamic acid decarboxylase immunopositive neurons in the rat medial preoptic area. *Neuroendocrinology* **48**, 591–602.

Levine, J.E. & Duffy, M.T. (1988). Simultaneous measurement of luteinizing hormone (LH)-releasing hormone, LH and follicle stimulating hormone release in intact and short-term castrated rats. *Endocrinology* **122**, 2211–2221.

Levine, J.E., Pau, K.-Y.F., Ramirez, V.D. & Jackson, G.L. (1982). Simultaneous measurement of luteinizing hormone-releasing and luteinizing hormone release in unanesthetized, ovariectomized sheep. *Endocrinology* **111**, 1449–1455.

Lincoln, D.W., Fraser, H.M., Lincoln, G.A., Martin, G.B. & McNeilly, A.S. (1985). Hypothalamic pulse generators. *Recent Progress in Hormone Research* **41**, 368–419.

Lincoln, G.A. & Short, R.V. (1980). Seasonal breeding: Nature's contraceptive. *Recent Progress in Hormone Research* **36**, 1–52.

MacLusky, N.J., Naftolin, F. & Leranth, C. (1988). Immunocytochemical evidence for direct synaptic connections between corticotropin-releasing factor (CRF)- and gonadotropin-releasing hormone (GnRH)-containing neurons in the preoptic area of the rat. *Brain Research* **439**, 391–395.

Maeda, K.-I., Tsukamura, H. & Yokoyama, A. (1987). Suppression of luteinizing hormone secretion is removed at late lactation in ovariectomized lactating rats. *Endocrinologia Japonica* **34**, 709–716.

Maeda, K.-I., Tsukamura, H., Ohkura, S., Kanaizuka, T. & Suzuki, J. (1991). Suppression of ovarian activity during the breeding season in suckling Japanese monkey (*Macaca fuscata*). *Journal of Reproduction and Fertility* **92**, 371–375.

Maeda, K.-I., Tsukamura, H., Uchida, E., Ohkura, N., Ohkura, S. & Yokoyama, A. (1989). Changes in the pulsatile secretion of LH after the removal of and subsequent resuckling by pups in ovariectomized lactating rats. *Journal of Endocrinology* **121**, 277–283.

Maeda, K.-I., Uchida, E., Tsukamura, H., Ohkura, N., Ohkura, S. & Yokoyama, A. (1990). Prolactin does not mediate the suppressive effect of the suckling stimulus on luteinizing hormone secretion in ovariectomized lactating rats. *Endocrinologia Japonica* **37**, 405–411.

McNeilly, A.S. (1988). Suckling and control of gonadotropin secretion. *The Physiology of Reproduction* (Knobil, E. and Neil, J.D., eds.), Raven Press, New York, pp. 2323–2349.

Merriam, R.G. & Wachter, K.W. (1982). Algorithms for the study of episodic hormone secretion. *American Journal of Physiology* **243**, E310–E318.

Mori, Y., Tanaka, M., Maeda, K., Hoshino, K. & Kano, Y. (1987). Photoperiodic modification of negative and positive feedback effects of oestradiol on LH secretion in ovariectomized goats. *Journal of Reproduction and Fertility* **80**, 523–529.

Ohkura, S., Tsukamura, H. & Maeda, K.-I. (1991). Effects of various types of hypothalamic deafferentation on luteinizing hormone pulses in ovariectomized rats. *Journal of Neuroendocrinology* **3**, 503–508.

Ohkura, S., Tsukamura, H. & Maeda, K.-I. (1992). Effects of transplants of fetal mediobasal hypothalamus on luteinizing hormone pulses impaired by hypothalamic deafferentation in adult ovariectomized rats. *Neuroendocrinology* **55**, in press.

Pelletier, G. (1987). Demonstration of contacts between neurons staining for LHRH in the preoptic area of the rat brain. *Neuroendocrinology* **46**, 457–459.

Silverman, A.J., Krey, L.C. & Zimmerman, E.A. (1979). A comparative study of the luteinizing hormone releasing hormone (LHRH) neuronal networks in mammals. *Biology of Reproduction* **20**, 98–110.

Smith, M.S., Lee, L.-R., Pohl, C.R., Lee, W.-S. & Hoffman, G.E. (1990). Inhibitory effects of the suckling stimulus on hypothalamic GnRH release and pulsatile LH secretion. *Neuroendocrine Regulation of Reproduction* (Yen, S.S.C. and Vale, W.W., eds.), Serono Symposia, Massachusetts, pp. 39–46.

Soper, B.D. & Weick, R.F. (1980). Hypothalamic and extrahypothalamic mediation of pulsatile discharges of luteinizing hormone in the ovariectomized rats. *Endocrinology* **106**, 348–355.

Stevenson, J.S. & Britt, J.H. (1981). Interval to estrus in sows and performance of pigs after alteration of litter size during late lactation. *Journal of Animal Science* **53**, 177–181.

Stevenson, J.S., Cox, N.M. & Britt, J.H. (1981). Role of the ovary in controlling luteinizing hormone, and prolactin secretion during and after lactation in pigs. *Biology of Reproduction* **24**, 341–353.

Taya, K. & Sasamoto, S. (1980). Initiation of follicular maturation and ovulation after removal of the litter from the lactating rat. *Journal of Endocrinology* **87**, 393–400.

Theodosis, D.T., Montagnese, C., Rodriguez, F., Vincent, J.-D. & Poulain, D.A. (1986). Oxytocin induces morphological plasticity in the adult hypothalamo-neurohypophysial system. *Nature* **322**, 738–740.

Tomogane, H., Ôta, K. & Yokoyama, A. (1969). Progesterone and 20α-hydroxypregn-4-en-3-one levels in ovarian vein blood of the rat throughout lactation. *Journal of Endocrinology* **44**, 101–106.

Tomogane, H., Ôta, K. & Yokoyama, A. (1976). Duration of diestrous period and secretion of progestins during prolonged lactation in the rat. *Endocrinologia Japonica* **23**, 137–141.

Tsukamura, H., Maeda, K.-I. & Yokoyama, A. (1988). Effect of the suckling stimulus on daily LH surges induced by chronic oestrogen treatment in ovariectomized lactating rats. *Journal of Endocrinology* **118**, 311–316.

Tsukamura, H., Maeda, K.-I., Ohkura, S. & Yokoyama, A. (1990). Effect of hypothalamic deafferentation on the pulsatile LH secretion of luteinizing hormone in ovariectomized lactating rats. *Journal of Neuroendocrinology* **2**, 59–63.

Tsukamura, H., Maeda, K.-I., Ohkura, S., Uchida, E. & Yokoyama, A. (1991). The suppressive effect of the suckling stimulus on the pulsatile luteinizing hormone release is not mediated by prolaction in the rat at mid-lactation. *Japanese Journal of Animal Reproduction* **37**, 59–63.

Weiner, R.I., Goldsmith, P.C., Windle, J.J. & Mellon, P.L. (1990). Cell biology and regulation of GnRH cell lines deprived from transgenic mice. *Neuroendocrinology* **52**(Suppl.), 21–22.

Wilson, R.C., Kesner, J.S., Kaufman, J.-M., Uemura, T., Akema, T. & Knobil, E. (1984). Central electrophysiological correlates of pulsatile luteinizing hormone secretion in the rhesus monkey. *Neuroendocrinology* **39**, 256–260.

Witkin, J.W., Paden, C.M. & Silverman, A.-J. (1982). The luteinizing hormone-releasing hormone (LHRH) systems in the rat brain. *Neuroendocrinology* **35**, 429–438.

Wuttke, W., Jarry, H., Demling, J., Wolf, R. & Duker, E. (1987). Involvement of GABA in the neuroendocrinology of reproduction. *Endocrinology and Physiology of Reproduction* (Leung, P.C.K., Armstrong, D.T., Ruf, K.B. Moger, W.H. and Friesen, H.G., eds.), Plenum Press, New York, pp. 65–69.

7 Brain Control of Female Sexual Behavior

Yasuo SAKUMA

Department of Physiology I, Hirosaki University School of Medicine, Hirosaki 036

The behavioral aspects of reproduction provide one of most distinctive examples of sex-related differences in brain activity. The implementation of female reproductive behavior in the rat depends critically on circulating estrogen. It is noteworthy that estrogen does not induce female-typical behavior when given to male rats orchidectomized as adults. Thus the difference in the behavioral expression between the sexes is not determined by the different gonadal hormones in the female and male; the brain apparently is responsible for sex-specific behaviors. The establishment of this sex difference of the brain also depends on gonadal hormones; like other somatic features, female or male pattern of the neural connections emerges in the brain at a specific stage of its development under the influence of gonadal hormones (MacLusky & Naftolin, 1981). Since the greatest amount of data has been collected about the hormonal control of the lordosis reflex, the principal component in the female rat reproductive behavior, this chapter will focus on brain mechanisms responsible for the display of this reflex under influences of circulating estrogen. Behavioral consequences of electrical stimulation, lesion or neural disruptions of specific brain sites will be discussed along with results of neuro-physiological recording studies to identify facilitatory and inhibitory neuronal circuitry for this behavior.

DESCRIPTION OF FEMALE SEXUAL BEHAVIOR

The copulatory encounter in the rat consists of several simple components which occur sequentially as results of interactions between the two sexes. In the females, precopulatory and copulatory phases can be distinguished; the former consists of soliciting or proceptive components, the latter is characterized by the display of lordosis reflex.

Proceptive Behavior

An increased locomotor activity in estrous female rats (Quadagno *et al.*, 1972) expedites their chance to run across males. When encountering a male, the female may show a characteristic gait, a hopping-darting sequence in the male's face, which is followed by an abrupt halt, usually assuming a crouching posture. This presents her hindquarters directly toward the male, allowing him to pursue and subsequently to mount her from the back. At this time or between each hopping-darting sequence, the highly receptive female often exhibits ear-wiggling, a rapid oscillatory movement of her head. These movements comprise solicitations, since they usually evoke mounting by the male partner (Pfaff *et al.*, 1973).

Estrogen is clearly responsible for the enhanced locomotor activity during the estrus, and its implants into the medial preoptic area (POA) of the ovariectomized female rats stimulate wheel-running (Fahrbach *et al.*, 1985). Swanson (1987) and his associates (1984) showed that POA efferents innervate the midbrain locomotor region, which centers in the caudal cuneiform nucleus, bordered medially by the midbrain central gray (CG) and extends caudally to the lateral parabrachial nucleus (Garcia-Rill, 1986). Neurons in the locomotor region are responsible for the activation of the spinal stepping mechanisms (Shik *et al.*, 1966). The CG and ventral tegmental area (VTA) are innervated by estrogen-accumulating POA neurons (Fahrbach *et al.*, 1986), but it is not known whether the POA projection to the locomotor region has affinity to the steroid. Therefore, it is too early to associate this neuronal projection with the initiation of sexual interactions in the rat, but this theory appears worthy of further investigation.

Lordosis Reflex

The consummatory phase of the female reproductive behavior in the rat as well as other rodents is characterized by the lordosis reflex, a postural reflex with a dorsiflexion of the vertebral column (Pfaff, 1980). Palpation of her flank, followed immediately by pressure stimuli on her perineum from the

Fig. 1.   A female rat showing lordosis in response to touch-pressure stimuli to her rump, tail-base and perineum given by male partner.

male partner, elicits this reflex. As a result of the spinal dorsiflexion and extension of the legs, the female elevates her head and rump, and reveals her perineal region (Fig. 1). This posture enables penile insertion by the male and insemination to occur. In the laboratory, similar cutaneous stimuli given manually by the experimenter elicits lordosis reflex in female rats in estrus. There are no other sensory requirements for the reflex, such as visual, olfactory or acoustic inputs, although some of them play important roles in more natural settings to initiate the proceptive components of the sexual interaction (Kow & Pfaff, 1976).

STEROID HORMONE ACTION ON THE BRAIN

The lack of lordosis reflex in cycling female rats on anestrous days or in the ovariectomized animals shows the estrogen-dependent nature of this reflex. Estrogen facilitates lordosis reflex when administered in specific brain sites. These include the POA, ventromedial nucleus of the hypothalamus (VMN) (Barfield & Chen, 1977) and medial amygdala (Lisk & Barfield, 1975). Neurons in these areas specifically accumulate estrogen (Pfaff & Keiner, 1972). Receptor molecules for estrogen in the neuronal nuclei in these structures are responsible for the uptake and retention of the steroid. The VMN is the site most sensitive to estrogen for the induction of lordosis reflex (Davis *et al.*, 1979). Electrical stimulation of this nucleus enhances the reflex in estrogen-treated, ovariectomized female rats, while its lesion disrupts it (Pfaff & Sakuma, 1979a, b). The CG mediates the facilitatory VMN effect on lordosis. Effects of both stimulation and lesion of the CG are evident in a much shorter latency than those of the VMN (Sakuma & Pfaff, 1979a, b).

The estrogen-sensitive VMN neurons with axons to the CG embody the neural substrate for this effect (Akaishi & Sakuma, 1986; Sakuma & Pfaff, 1982; Sakuma, 1984). Estrogen increases the excitability of these neurons when driven antidromically from the CG (Sakuma, 1984; Sakuma & Akaishi, 1987); conversely, estrogen suppresses POA neurons with axons to the VTA, which are presumed to inhibit lordosis (Hasegawa & Sakuma, 1991). The nature of the estrogen-induced excitability change is not clear at present. Besides the classic genomic action, estrogen may act at the neuronal membrane in some receptor-mediated fashion, or modulate the interaction between neurotransmitters with their receptors, by causing, for instance, changes in the membrane fluidity (Erulkar & Wetzel, 1987).

Neuronal excitability is determined either intrinsically by voltage-dependent membrane conductance or extrinsically by neurotransmitters (Llinas, 1988). Schiess *et al.*, (1988) proposed that chronic estrogen-treatment may affect the depolarizing process of the medial amygdala neurons through regulation of the calcium-dependent K^+ channel, a determinant of the intrinsic property of the membrane. Prior treatment with estrogen also augmented adrenergic depolarization in the guinea pig arcuate neurons in slices (Condon *et al.*, 1989). A recent patch-clamp study has shown in the myotubes of *Xenopus* laryngeal muscle that incubation in the steroid-containing medium increases acetylcholine-activated, single-channel conductance along with non-specific changes in single-channel kinetics (Erulkar & Wetzel, 1989). The membrane action of steroid hormones is not surprising since these hormones share the cyclopentanoperhydrophenantrane nucleus with the cardiac glycosides, which combine at the myocardial cell surface and inhibit Na^+-K^+ pump. Tetrahydroprogesterone, an A-ring reduced metabolite of progesterone, acts as an agonist for $GABA_A$ receptors (Majewska *et al.*, 1986), while pregnenolone sulfate has allosteric antagonistic activity to the same receptor (Majewska *et al.*, 1988). Glial cells in the brain contain enzymes producing these and related "neurosteroids" (Jung-Testas & Hu, 1989; Majewska *et al.*, 1988) from cholesterol and other precursors, which may act locally, together with those produced by gonads and carried into the brain by the systemic circulation.

Different time courses, but not its direction, distinguish the two routes of acute estrogen action on the neuronal excitability. Following addition to the perfusion medium, estrogen depolarizes hypothalamic arcuate neurons (Kelly *et al.*, 1980) or hyperpolarizes amygdala neurons at short latencies (Nabekura *et al.*, 1986). The inhibition of amygdala neurons was caused by an induction of K^+ efflux, independent of protein synthesis (Nabekura *et al.*, 1986).

On the other hand, testosterone, which is readily converted to estrogen in the brain (Kendrick & Drewett, 1980; MacLusky & Naftolin, 1981), needs 48–72 hr to excite POA efferents to the medial forebrain bundle and to reinstate sexual behavior in the castrated male rat (Kendrick *et al.*, 1981). Similarly, in the female rat, estrogen shortened the refractory period and lowered the antidromic activation threshold at about 48–72 hr in the oxytocinergic neurons in the paraventricular-neurohypophyseal system (Akaishi & Sakuma, 1985), or in the VMN neurons with axons to the CG (Sakuma & Pfaff, 1982; Sakuma, 1984).

In both male and female rats, the long latency for the neuronal activation coincided with the time needed for the restoration of sexual responses by respective hormones, which can be blocked by inhibitors of protein synthesis (Parsons *et al.*, 1982). These and other observations suggest that some of the estrogen-induced, long-lasting changes in electrical excitability in neurons and uterine muscles are mediated by genomic activation (McEwen, 1991). Other examples include the diminished accommodation in the amygdala neurons following chronic estrogen, that usually follows repetitive generation of action potentials by intracellular current injection (Schiess *et al.*, 1988). Indeed, the long-term estrogen treatment induces K^+ channel mRNA in rat uterus, which generates a voltage-dependent, slow K^+ current when expressed in *Xenopus* oocytes (Pragnell *et al.*, 1990). The slow activation of this current may relate to the loss of the accommodation in the amygdala neurons, by allowing the bursting action potentials to continue for several seconds before their termination by K^+ efflux.

It has been pointed out that long-term estrogen lowers thresholds for antidromic activation from the midbrain in a subset of VMN neurons with axons in the supraoptic commissure, but not in the dorsal longitudinal fascicle (Akaishi & Sakuma, 1986; Sakuma & Akaishi, 1987). In the medial basal hypothalamus, the location of the estrogen-sensitive neurons agrees with those of the anatomically demonstrated estradiol-concentrating cells, which presumably contain the nuclear receptor molecule for this steroid; about 30% of the neurons respond to microiontophoretically infused estradiol (Kelly, 1882), which corresponds to the percentage of the estradiol-concentrating cells in this region (Morrell *et al.*, 1986). Therefore, estrogen appears to act on identical, specific populations of neurons to cause different responses at varying latencies. Possible mechanisms for estrogen action on neuronal excitability may include the initiation of the synthesis and insertion of new ion channels into the membrane (which obviously needs time) and later modulation of such channels by itself.

HORMONAL REGULATION OF FEMALE SEXUAL BEHAVIOR

Lordosis reflex can be reinstated within 12–48 hr in ovariectomized female rats following estrogen treatment. Systemic progesterone supplement at this time further potentiates the reflex at about 6 hr after the injection. It should be emphasized that estrogen alone suffices to induce lordosis reflex in the ovariectomized female rats; progesterone complements the expression of the proceptive behavior. Estrogen in a large amount, or other compounds such as blockers of neural transmission or certain hypothalamic peptides, replaces this progesterone action. Whereas the latency for progesterone action can be shortened to 10–60 min by applying intravenously or by direct implants into the brain (McEwen, 1988), the time needed for estrogen action does not depend on the route of administration (Parsons *et al.*, 1982).

Some of the progesterone action may be manifested on the neuronal membrane, since membrane-active metabolites of this steroid enhance lordosis reflex when given to the medial POA (Rodriguez-Manzo *et al.*, 1986). Progesterone or its metabolites may either directly modulate ion flux across neuronal membrane or activate the intracellular messenger system. Cyclic AMP is one of several candidates for the intracellular mediation of the progesterone effect. Cyclic AMP (Beyer *et al.*, 1981) and forskolin (Mobbs *et al.*, 1989), an adenyl cyclase activator, enhance the lordosis reflex in ovariectomized, estrogen-treated female rats. Cyclic AMP also mimics acute depolarizing action of estrogen on neurons in the VMN nucleus; the estrogen-effect is augmented by forskolin (Minami *et al.*, 1990). This may relate to an observation that estrogen in large doses can replace progesterone to enhance lordosis reflex, presumably *via* progesterone receptor (Parsons *et al.*, 1984). Progesterone may also enhance the reflex *via* its effect on lutenizing hormone-releasing hormone (LHRH) release from the presynaptic terminals, as has been shown to occur in the hypothalamic slices (Ke & Ramirez, 1987). LHRH (Sakuma & Pfaff, 1980a) or its C-terminal fragment (Dudley & Moss, 1988) promotes lordosis reflex when infused into the POA, VMN or the CG. LHRH may act by raising cyclic AMP levels in the brain, since systemic methylxantines, which inhibit the catalytic effect of phosphodiesterase on cyclic AMP, significantly synergized with LH-RH to facilitate lordosis reflex (Beyer *et al.*, 1982).

Glucocorticoids antagonize estrogen and suppress lordosis reflex (de Catanzaro *et al.*, 1985). Potential site for this antagonism appears to be the hippocampus. Lesions in hippocampal areas CA 1 and 2, the site of neurons with glucocorticoid receptor, enhance lordosis reflex (Cameron *et al.*, 1979). Glucocorticoid activates hippocampal neurons (Pfaff *et al.*, 1971), whereas estrogen, when given chronically, hyperpolarizes pyramidal neurons in this region (Beck *et al.*, 1989). Several neurotransmitters are implicated in the hippocampal responses to estrogen and glucocorticoids. In the female rat, estrogen augments serotonergic effects through its 1A receptor to produce hyperpolarization (Clarke & Goldfarb, 1989; Clarke & Maayani, 1990). Elimination of serotonergic innervation by hippocampal infusion of 5,7-dihydroxytryptamine causes excitation of local neurons, which is reversed by local raphe grafts (Richter-Levin & Segal, 1990). However, in the light of observation that hippocampal infusion of methysergide, a serotonergic blocker, promotes lordosis reflex (Franck & Ward, 1981), the synergism between estrogen and serotonin in the hippocampus may not be related to the control of the reflex. If this is the case, steroidal effects on other transmitter systems should be taken into account. One candidate is GABA, since $GABA_A$ receptor is induced in response to estrogen or adrenalectomy, while corticosterone has the opposite effect (Kendall *et al.*, 1982). GABAergic facilitation of the lordosis reflex in the medial hypothalamus is suggested based on the effects of the local application of bicuculline (McCarthy *et al.*, 1990). Although no direct evidence is available for hippocampal GABAergic action on the lordosis reflex, the facilitatory effect of local progesterone implant in the hippocampus (Franck & Ward, 1981) can be ascribed to its agonistic action on $GABA_A$ receptor (Majewska *et al.*, 1986).

The long latency for the estrogen action, together with its effective interruption by blockers of the intracellular protein synthesis, suggests the genomic mediation of the estrogen effect to induce the lordosis reflex. Indeed, estrogen increased the degree of stacking of the rough endoplasmic reticulum in VMN neurons (Meisel & Pfaff, 1985) and the number of dense-cored vesicles in presumed VMN terminals in the CG (Chung *et al.*, 1988), both of which were positively related to the level of lordosis reflex. An increased level of progesterone receptor protein was detected at about 12 hr following estrogen treatment, simultaneously with the behavioral activation (McEwen *et al.*, 1979). Other protein molecules, including phospholipase isozyme (Mobbs *et al.*, 1990a, b), oxytocin receptor (De Kloet *et al.*, 1986), muscarinic receptor (Rainbow *et al.*, 1984), prolactin (Shivers *et al.*, 1989) and proopiomelanocortin (Romano *et al.*, 1990; Tong *et al.*, 1990), are induced in the brain by estrogen or combined treatments with estrogen and progesterone. On the other hand, estrogen suppresses brain expression of cholecystokinin (Akesson *et al.*, 1987) or $GABA_A$ receptor (Canonaco *et al.*, 1989). Prolactin has been shown to enhance lordosis reflex when infused into the CG (Harlan *et al.*, 1983), and some ligands for these receptors (*e.g.*, oxytocin) have been suggested to interact with its receptor in the dendritic field of the VMN neurons to promote the reflex (Schumacher *et al.*, 1990).

NEURAL CIRCUITRY FOR THE FEMALE RAT SEXUAL BEHAVIOR

Facilitatory Brain Control of Lordosis

Based on studies employing stereotaxic implants of estradiol, focal electrical stimulation or lesions, neurons in the VMN are now generally accepted as the essential site of estrogen action to promote the lordosis reflex in the rat and other rodents (Pfaff & Schwartz-Giblin, 1988). Implantation of the fetal brain tissue containing the VMN compensates lesion effects in the adult females (Jenssen *et al.*, 1988). Behavioral (Malsbury & Daood, 1978; Manogue *et al.*, 1980) and electrophysiological recording (Akaishi & Sakuma, 1986) studies in animals with hypothalamic knife cuts revealed that the axons leaving the VMN in the anterolateral direction, which eventually terminate in the CG, are essential for the induction of this reflex. We have shown that estrogen increases the electrical excitability of neurons in this system in close association with the induction of lordosis reflex. Other CG efferents of the VMN, which descend in the periventricular stratum, are neither sensitive to the steroid nor required for the behavioral activation. The estrogen-sensitive efferents of the VMN activate neurons in the dorsal subdivision of the CG (Sakuma & Pfaff, 1980b), which, in turn, projects to the medullary reticular field (Sakuma & Pfaff, 1980c). Involvement of the reticulospinal and vestibulospinal neurons in the display of lordosis reflex, through the excitation of motoneurons for the axial musculature, has been demonstrated (Pfaff & Schwartz-Giblin, 1988).

In estrous female rats, sensory afferents from the pressure receptors in the skin of the rump, tail-base and perineum region, which ascend in the spinal anterolateral column, are needed to trigger lordosis reflex (Kow *et al.*, 1977). Hypothalamic responses to these sensory inputs are unlikely to be important, since, in animals under light urethane anesthesia, hypothalamic neurons do not respond specifically and promptly to these stimuli. The baseline rate of spontaneous activity in VMN neurons is also very low in the estrogen-treated animals for regulating lordosis reflex, since the reflex occurs at about 200 msec following the initial sensory stimuli from the male partner (Pfaff, 1980). Autoradiography of 2-deoxyglucose uptake in the awake, free moving vole revealed that lordosis reflex is accompanied by neuronal activation in the CG, but not in the VMN and other hypothalamic nuclei (Sutherland *et al.*, 1984.). Behavioral evidence also suggests that the hypothalamus is not included in the reflex arc for the lordosis. The disruptive effects of lesions in the VMN on the lordosis reflex take hours to appear (Pfaff & Sakuma, 1979b); in contrast, those in the CG have an immediate effect (Sakuma & Pfaff, 1979b). These observations indicate that an important feature of the hypothalamic involvement in the lordosis control is its estrogen dependence. The estrogen-sensitive efferents regulate the transmission in the reflex arc, which apparently lies below the brainstem (Robbins *et al.*, 1990).

Inhibitory Neurons for Lordosis

Lesions in the lateral septum increases behavioral sensitivity of ovariectomized female rats to estrogen (Lisk & MacGregor, 1982; Nance *et al.*, 1975). In addition, electrical stimulation of this structure disrupts lordosis reflex in the rat (Moss *et al.*, 1974), as well as in the hamster (Zasorin *et al.*, 1975). Thus the estrogen-accumulating neurons in the lateral septum are implicated as an origin of the lordosis-inhibiting impulses. Some axons of these neurons penetrate the POA on its way to the brainstem (Swanson, 1987), and stereotaxic interruption of this pathway by dorsal deafferentation of the POA not only enhanced lordosis reflex in female rats (Yamanouchi & Arai, 1977, 1990) but also induced this behavior in male (Yamanouchi & Arai, 1978) or androgenized female rats (Kondo *et al.*, 1986).

It has been shown that the behavioral activation by the septal lesion operates independent of the hypothalamic facilitatory mechanism for the lordosis reflex. A knife cut placed anterodorsally to the VMN, which presumably eliminate septal efferent to the medial basal hypothalamus, does not interfere with the enhanced lordosis in the septum-lesioned rat (King & Nance, 1985). On the other hand, an additional lesion of the medial amygdala effectively antagonizes the behavioral effect of the septal lesion; animals with both septal and amygdala lesions have lower lordosis-to-mount ratio than those with lesions confined to the septum (McGinnis *et al.*, 1978). Lesions of the medial amygdala, which also contains estrogen-sensitive cells (Pfaff & Keiner, 1972), disrupt lordosis reflex while its electrochemical stimulation activates the reflex (Masco & Carrer, 1980). Amygdala implants of estrogen induce the reflex when supplemented with systemic progesterone (Lisk & Barfield, 1975). Thus, the enhanced estrogen sensitivity in septum-lesioned rats may be mediated by the medial amygdala.

The POA is another estrogen-sensitive structure which attracted extensive scrutiny for its role in the regulation of the lordosis reflex. The POA lesion causes an increased behavioral sensitivity (Powers & Valenstein, 1972) and its stimulation disrupts the lordosis reflex (Moss *et al.*, 1974;

Pfaff & Sakuma, 1979a). However, the complex anatomy and function of this structure occasionally caused confusions in the interpretation of these data. This area contains fibers of passage originating in the septum (Swanson, 1987). This led to speculation that the increased behavioral sensitivity produced by the POA lesion was the result of interference in the fiber of passage rather than the destruction of local cell bodies (Clough *et al.*, 1986; Yamanouchi & Arai, 1977). Some lesions of the POA diminish lordosis reflex and ovulation altogether, presumably by obliterating descending LH-RH projections (Bast *et al.*, 1987; Leedy, 1984; Popolow *et al.*, 1981), which otherwise promote lordosis reflex. The absence of the effect of the prepubertal lesion in the non-ovariectomized animals might be the result of

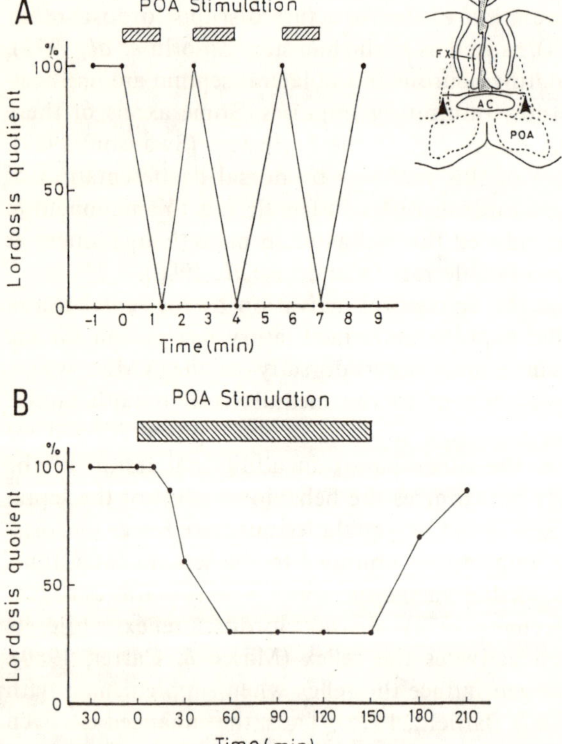

Fig. 2.   Examples of lordosis reflex suppression during bilateral electrical stimulation of the preoptic area (POA) in the transected (A) and intact (B) animals. Bilateral stimulation was given in 30-sec trains for periods indicated by the bar in each panel, with parameters: stimulus frequency 100 Hz, 60 μA per electrode. Inset shows a representative frontal section of the brain with the transection (arrow heads), immediately dorsal to the the anterior commissure (AC). FX, fornix.

functional compensation at the time of behavioral observation several weeks after the surgery, which is evidenced by the reinstatement of the estrous cyclicity (Clough *et al.*, 1986). Transplantation of POA grafts containing LH-RH neurons resuscitates lordosis reflex in the hypogonadal mice (Gibson *et al.*, 1987). In addition, the initiation of the sexual interaction depends on the integrity of the POA. For example, female rats with POA lesions avoid sexual contacts with male partners (Whitney, 1986) and hamsters bearing these lesions fail to express proceptive behaviors (Floody, 1989).

We showed recently in animals deprived of POA afferents from the septum by knife-cuts that, notwithstanding the removal of septal inhibition, electrical stimulation causes a prompt and strong suppression of lordosis reflex (Fig. 2) (Takeo *et al.*, 1992). Therefore, the POA appears to be an independent and separate entity in the inhibitory mechanism of the reflex, in addition to septal or other limbic structures. As a corollary, ascending POA projection into the septum (Conrad & Pfaff, 1976; Simerly & Swanson, 1988), which originates from cells in the ventral part of the POA (Luiten *et al.*, 1982), is not involved in the suppression of the reflex.

Our results suggest that estrogen-sensitive projection of the POA to the VTA (Hasegawa & Sakuma, 1990) is the neural substrate for the POA suppression of the lordosis. Estrogen alters the excitability of VTA afferents of the POA origin in a sexually dimorphic manner, in close association with the behavioral sensitivity of each animal to this hormone (Hasegawa

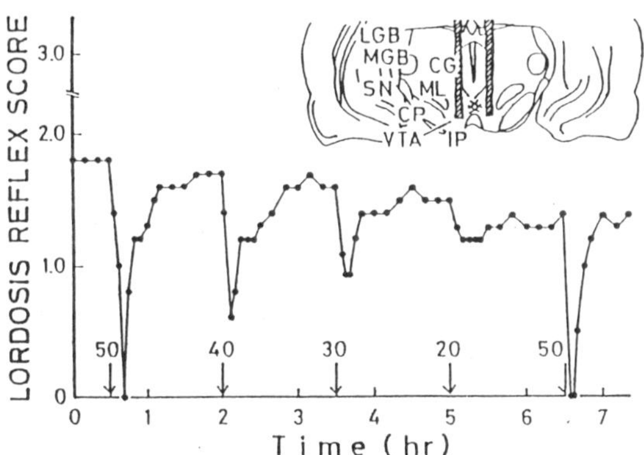

Fig. 3.   Inhibition of lordosis reflex from the ventral tegmental area (VTA) in free-moving, ovariectomized, estrogen-treated rat. Each stimulus (100 Hz pulse train) was delivered bilaterally at points indicated by arrows. Number gives current intensity (μA). Abbreviations in the inset: CG, midbrain central gray; CP, cerebral peduncle; LGB and MGB, lateral and medial geniculate bodies; ML, medial lemniscus, SN, substantia nigra.

& Sakuma, 1992). Electrical stimulation of the VTA causes intense and prompt disruption of lordosis reflex in estrogen-treated ovariectomized female rats, reminiscence of the results obtained from the rats bearing the deafferented POA (Fig.3) (Hasegawa *et al.*, 1991). Estrogen may decrease the neuronal activity in the VTA by reducing POA tonus. The net effect of neurons in the POA appears to be an inhibition of the lordosis reflex, with the obvious exception of LH-RH cells scattered in this region (Pfaff & Schwartz-Giblin, 1988).

The VTA-induced interruption of lordosis does not require the integrity of the dopaminergic system. Therefore, we suggest involvement of non-dopaminergic efferents of this structure to the midbrain or pontine structures (Fallon *et al.*, 1984; Kalen *et al.*, 1988). VTA axons descend in the ventral part of the CG and extend as far caudal as the dorsal pontine tegmentum to the level of locus coeruleus (Beckstead *et al.*, 1979; Swanson, 1982). Electrical stimulation of the ventral CG, contrary to that of the dorsal subdivision (Sakuma & Pfaff, 1980), causes a significant disruption of lordosis (Arendash & Gorski, 1983). Functional dichotomy between the ventral and dorsal subdivisions of the CG is also noted in the autonomic regulation (Lovick, 1991) or pain control (Cannon *et al.*, 1982).

Electrical stimulation (Sakai, 1990) or local infusion of carbachol (Hayes *et al.*, 1984) to the dorsal pontine tegmentum produces postural atonia due to a generalized inhibition of spinal motoneurons (Sakai, 1990), an effect presumably implemented by activation of reticulospinal neurons in the medullary inhibitory center (Jankowska *et al.*, 1968). Such effect would suffice to dissolve the lordosis posture in the female rat. It is of interest that the POA (Kaitin, 1984) as well as the dorsal pontine tegmentum (Sakai, 1990) are implicated in the induction of the paradoxical sleep, which accompanies postural atonia. This phase of sleep is now usually considered to be equivalent to the "electroencephalographic after-reaction" in the postovulatory animals (Sawyer & Kawakami, 1959), which presumably follows a decline in the circulating level of estrogen.

Brain Control on the Initiation of Sexual Interaction
Whereas activation of the estrogen-sensitive POA efferents to the VTA suppresses reflexive display of the lordosis, the POA appears to initiate sexual interactions and several other motivational behaviors. POA lesions compel female rats to avoid contact with males, despite their enhanced ability to show the reflex (Whitney, 1986). In female monkeys in estrus, a transitory activation of some POA neurons precedes the commencement of reproductive interaction with the males (Aou *et al.*, 1988).

Sexual odors play important roles in the initiation of copulatory interaction in several species (Whitten & Champlin, 1973). The effect appears to depend on the integrity of the accessory olfactory system in the rat; removal of the vomeronasal organ (Rajendren *et al.*, 1990) or electrolytic lesion in the corticomedial nucleus of the amygdala reduces the probability of lordosis occurrence, while electrochemical stimulation of the latter has the opposite effect (Masco & Carrer, 1980). The accessory olfactory inputs arrive at the POA and medial hypothalamus (Scalia & Winans, 1975) by way of the medial amygdala as a component of the stria terminalis (De Olmos & Ingram, 1971; Leonard & Scott, 1971). The effect may be due to the loss of olfactory cues for the initiation of copulatory interaction. This hypothesis needs scrutiny since the amygdala also may affect reflexive implementation of lordosis through its projection to the VMN (Renaud, 1976), which plays a key role in the display of the lordosis.

SEXUAL DIMORPHISM IN THE SEX-HORMONE SENSITIVE NEURAL CIRCUITRY

Morphological sex differences in the brain are noted in the synaptic as well as neuronal organizations of the arcuate nucleus (Arai & Matsumoto, 1978), the VMN (Matsumoto & Arai, 1986), medial amygdala (Nishizuka & Arai, 1981) and the POA (Raisman & Field, 1973; Gorski, 1984). Sexually dimorphic synaptic convergence in the POA is also evidenced by electrophysiological recordings (Dyer *et al.*, 1976). Sex differences are also found at the level of neuronal plasma membrane, in the number of intramembrane protein particles or Concanavalin binding sites in the arcuate nucleus (Garcia-Segura *et al.*, 1988, 1989). Sex difference in autonomic, endocrine and behavioral functions, which, as a matter of course, include sexual behavior, may be the result of these sexually dimorphic features (Arnold & Gorski, 1984).

The influence of the neonatal endocrine environment on the establishment of the sex difference in reproductive and other brain functions in rodents is well documented (MacLusky & Naftolin, 1981). Electrophysiological properties of the efferents of the VMN and the POA are also determined by the neonatal treatment of testosterone (Sakuma, 1984; Hasegawa & Sakuma, 1992). In parallel with the display of lordosis, these neurons respond to estrogen in ovariectomized female rats or neonatally orchidectomized male rats, but not in females neonatally exposed to testosterone. It is noteworthy that the VMN and POA neurons respond to estrogen in opposite ways; when effective, estrogen decreased the threshold and shortened the refractory period for antidromic activation of VMN neurons from the CG (Sakuma, 1984), whereas it had inverse effects on

POA neurons with axons to the VTA (Hasegawa & Sakuma, 1990; 1992). This is an additional evidence that the lordosis reflex is under the regulation of dual, antagonistic mechanisms. In order to implement this reflex, estrogen activates the VMN and disinhibits the POA; the lack of estrogen-sensitivity in both structures in neonatally testosterone-exposed rats prevents a display of the lordosis reflex.

It remains to be seen how the estrogen-sensitive mechanism operates in individual neurons in the females and neonatally orchidectomized males or how neonatal testosterone invalidates it. One theory explaining the lack of estrogen effect in the testosterone-exposed females may be the elimination of the subset of the neuronal population with this particular mechanism. Regressive events in neurogenesis, as a result of competition among neurons produced in excess of adult needs, have been widely accepted as a distinct phase of the development of the vertebrate nervous system (Cowan *et al.*, 1984). A selective promotion of growth (Toran-Allerand, 1976) or death of a certain class of neurons by neonatal exposure to testosterone or other hormones in blood (Cowan *et al.*, 1984) would lead to the establishment of an adult brain with different characteristics.

CONCLUSIONS

Since the lordosis reflex was first described in 1940's as the principal component in the female rat reproductive behavior, our understanding of the control of this reflex has increased enormously. This applies to both neural and hormonal aspects, for we now have almost complete schema of the neuronal circuitry for the lordosis reflex, and know where and how estrogen modulates its workings (Fig. 4). Although earlier contributions were made by behavioral scientists, this reflex has attracted investigators from many different disciplines in neuroscience lately, because it provides an ideal model for steroid action on the brain, which is free from pituitary intervention and therefore completely visible to the investigator. The recent interest in the role of steroid hormones in brain development not only provides one example of the physiological principles which have been established during the course of investigation of the lordosis reflex, but also has far wider potential.

As more answers are provided by neurophysiological studies on this reflex, more questions are generated. Perhaps the most urgent among them relates to the mechanism of rapid effects of steroid hormones on the neuronal activity (Schumacher, 1990), which presumably varies between the sexes or individual brain sites. In longer terms, steroid effects on higher nervous activity such as memory, which is exemplified by song acquisition

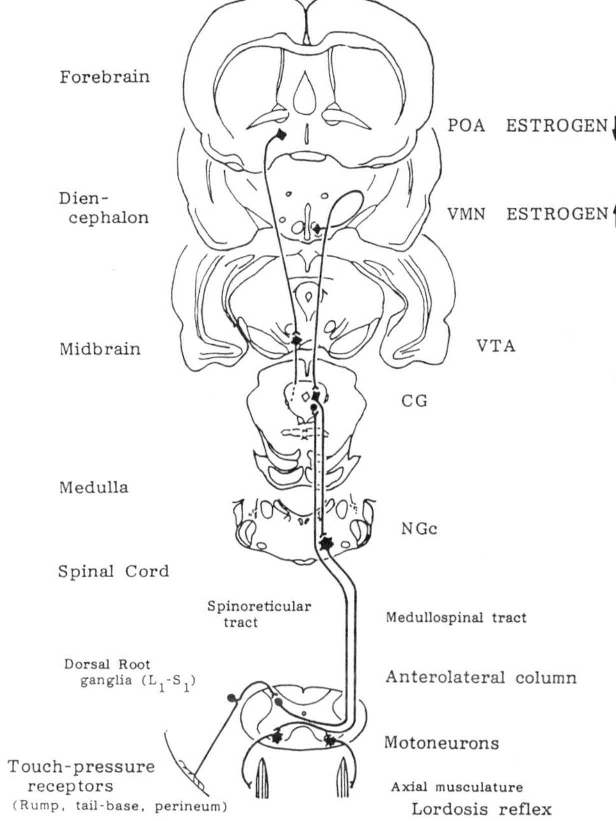

Fig. 4.   A model depicting the neuronal circuitry that may relevant for lordosis reflex in the rat. NGc, gigantcellular nucleus of the medulla.

in birds (Marler, 1991), pose an even more intriguing question toward understanding the interactions of steroid hormone and brain.

REFERENCES

Akaishi, T. & Sakuma, Y. (1985). Estrogen excites oxytocinergic, but not vasopressinergic cells in the paraventricular nucleus of female rat hypothalamus. *Brain Research* **335**, 302–305.

Akaishi, T. & Sakuma, Y. (1986). Projection of oestrogen-sensitive neurones from the ventromedial hypothalamic nucleus of the female rat. *Journal of Physiology* **372**, 207–220.

Akesson, T., Mantyh, P., Mantyh, C., Myatt, D. & Micevych, P. (1987). Estrous cyclicity of ^{125}I-cholecystokinin octapeptide binding in the ventromedial hypothalamic nucleus. *Neuroendocrinology* **45**, 257–262.

Aou, S., Oomura, Y. & Yoshimatsu, H. (1988). Neuron activity of the ventromedial hypothalamus and the medial preoptic area of the female monkey during sexual behavior. *Brain Research* **455**, 65–71.

Arai, Y. & Matsumoto, A. (1978). Synapse formation of the hypothalamic arcuate nucleus during post-natal development in the female rat and its modification by neonatal estrogen treatment. *Psychoneuroendocrinology* **3**, 31–45.

Arendash, G.W. & Gorski, R.A. (1983). Suppression of lordotic responsiveness in the female rat during mesencephalic electrical stimulation. *Pharmacology, Biochemistry and Behavior* **19**, 351–357.

Arnold, A.P. & Gorski, R.A. (1984). Gonadal steroid induction of structural sex differences in the central nervous system. *Annual Review of Neuroscience* **7**, 413–442.

Barfield, R.J. & Chen, J.J. (1977). Activation of estrous behavior in ovariectomized rats by intracerebral implants of estradiol benzoate. *Endocrinology* **101**, 1716–1725.

Bast, J.D., Hunts, C., Renner, K.J., Morris, R.K. & Quadagno, D.M. (1987). Lesions in the preoptic area suppressed sexual receptivity in ovariectomized rats with estrogen implants in the ventromedial hypothalamus. *Brain Research Bulletin* **18**, 153–158.

Beck, S.G., Clarke, W.P. & Goldfarb, J. (1989). Chronic estrogen effects on 5-hydroxytryptamine mediated responses in hippocampal pyramidal cells of female rats. *Neuroscience Letters* **106**, 181–187.

Beckstead, R.M., Domesick, V.B. & Nauta, W.J. (1979). Efferent connections of the substantia nigra and ventral tegmental area in the rat. *Brain Research* **175**, 191–217.

Beyer, C., Canchola, E. & Larsson, K. (1981). Facilitation of lordosis behavior in the ovariectomized estrogen primed rat by dibutyryl cAMP. *Physiology and Behavior* **26**, 249–251.

Beyer, C., Gomora, P., Canchola, E. & Sandval, Y. (1982). Pharmacological evidence that LH-RH action on lordosis behavior is mediated through a rise in c-AMP. *Hormones and Behavior* **16**, 107–112.

Cameron, W.R., Gage, F.H., III, Hitt, J.C. & Popolow, H.B. (1979). The influence of the anterodorsal hippocampus on feminine sexual behavior in rats. *Behavior and Neural Biology* **27**, 72–86.

Cannon, J.T., Prieto, G.J., Lee, A. & Liebeskind, J.C. (1982). Evidence for opioid and non-opioid forms of stimulation-produced analgesia in the rat. *Brain Research* **243**, 315–312.

Canonaco, M., O'Connor, L.H., Pfaff, D.W. & McEwen, B.S. (1989). Longer term progesterone treatment induces changes of $GABA_A$ receptor levels in forebrain sites in the female hamster: quantitative autoradiography study. *Experimental Brain Research* **77**, 407–411.

Chung, S.K., Pfaff, D.W. & Cohen, R.S. (1988). Estrogen-induced alterations in synaptic morphology in the midbrain central gray. *Experimental Brain Research* **69**, 522–530.

Clarke, W.P. & Goldfarb, J. (1989). Estrogen enhances a 5-HT$_{1A}$ response in hippocampal slices from female rats. *European Journal of Pharmacology* **160**, 195–197.

Clarke, W.P. & Maayani, S. (1990). Estrogen effects on 5-HT$_{1A}$ receptors in hippocampal membranes from ovariectomized rats: functional and binding studies. *Brain Research* **518**, 287–291.

Clough, R.W., Rodriguez-Sierra, J.F. & Hoppe, L.B. (1986). Kainic acid lesioning of the preoptic area alters positive feedback of estrogen in prepubertal female rats. *Biology of Reproduction* **35**, 1269-1276.

Condon, T.P., Ronnekleiv, O.K. & Kelly, M.J. (1989). Estrogen modulation of the α-1-adrenergic response of hypothalamic neurons. *Neuroendocrinology* **50**, 51-58.

Conrad, L.C.A. & Pfaff, D.W. (1976). Efferents from medial basal forebrain and hypothalamus in the rat. I. An autoradiographic study of the medial preoptic area. *Journal of Comparative Neurology* **169**, 185-220.

Cowan, W.M., Fawcett, J.W., O'Leary, D.D.M. & Stanfield, B.B. (1984). Regressive events in neurogenesis. *Science* **225**, 1258-1265.

Davis, P.G., McEwen, B.S. & Pfaff, D.W. (1979). Localized behavioral effects of tritiated estradiol implants in the ventromedial hypothalamus of female rats. *Endocrinology* **104**, 898-903.

De Catanzaro, D., Lee, P.C.S. & Kerr, T.H. (1985). Facilitation of sexual receptivity in female mice through blockade of adrenal 11β-hydroxylase. *Hormones and Behavior* **19**, 77-85.

De Kloet, E.R., Voorhuis, D.A.M., Boschma, Y. & Elands, J. (1986). Estradiol modulates density of putative "oxytocin receptors" in discrete rat brain regions. *Neuroendocrinology* **44**, 415-421.

De Olmos, J.S. & Ingram, W.R. (1972). The projection field of the stria terminalis in the rat brain. An experimental study. *Journal of Comparative Neurology* **146**, 303-334.

Dudley, C.A. & Moss, R.L. (1988). Facilitation of lordosis in female rats by CNS-site specific infusions of an LH-RH fragment, Ac-LH-RH$^{5-10}$. *Brain Research* **441**, 161-167.

Dyer, R.G., MacLeod, N.K. & Ellendorff, F. (1976). Electrophysiological evidence for sexual dimorphism and synaptic convergence in the preoptic and anterior hypothalamic area of the rat. *Proceedings of the Royal Society, Series B* **193**, 421-440.

Erulkar, S.D. & Wetzel, D.M. (1987). Steroid effects on excitable membranes. *Current Topics in Membrane and Transport* **31**, 141-190.

Erulkar, S.D. & Wetzel, D.M. (1989). 5α-Dihydrotestosterone has nonspecific effects on membrane channels and possible genomic effects on ACh-activated channels. *Journal of Neurophysiology* **61**, 1036-1052.

Fahrbach, S.E., Meisel R.L. & Pfaff D.W. (1985). Preoptic implants of estradiol increase wheel running but not the open field activity of female rats. *Physiology and Behavior* **35**, 985-992.

Fahrbach, S.E., Morrell, J.I. & Pfaff, D.W. (1986). Identification of medial preoptic neurons that concentrate estradiol and project to the midbrain in the rat. *Journal of Comparative Neurology* **247**, 364-382.

Fallon, J.H., Schmued, L.C., Wang, C., Miller, R. & Banales, G. (1984). Neurons in the ventral tegmentum have separate populations projecting to telencephalon and inferior olive, are histochemically different, and may receive direct visual input. *Brain Research* **321**, 332-336.

Femano, P.A., Schwarz-Giblin, S. & Pfaff, D.W. (1984). Brain stem reticular influences on lumbar axial muscle activity. I. Effective sites. *American Journal of Physiology* **15**, R389-R395.

Floody, O.R. (1989). Dissociation of hypothalamic effects on ultrasound production and copulation. *Physiology and Behavior* **46**, 299-307.

Franck, J.A.E. & Ward, I.L. (1981). Intralimbic progesterone and methysergide facilitate lordotic behavior in estrogen-primed female rats. *Neuroendocrinology* **32**, 50-56.

Garcia-Rill, E. (1986). The basal ganglia and the locomotor regions. *Brain Research Review* **11**, 47–63.

Garcia-Segura, L.M., Perez, J., Tranque, P.A., Olmos, G. & Naftolin, F. (1988). Sexual differentiation of the neuronal plasma membrane: neonatal levels of sex steroids modulate the number of exo-endocytotic images in the developing rat arcuate neurons. *Neuroscience Letters* **91**, 19–23.

Garcia-Segura, L.M., Perez, J., Tranque, P.A., Olmos, G. & Naftolin, F. (1989). Sex differences in plasma membrane Concanavalin A binding in the rat arcuate neurons. *Brain Research Bulletin* **22**, 651–665.

Gibson, M.J., Moscovitz, H.C., Kokoris, G.J. & Silverman, A.-J. (1987). Female sexual behavior in hypogonadal mice with GnRH-containing brain grafts. *Hormones and Behavior* **21**, 211–222.

Gorski, R.A. (1984). Critical role for the medial preoptic area in the sexual differentiation of the brain. In: *Progress in Brain Research*, Vol. 61, Sex Differences in the Brain: Relation between Structure and Function (De Vries, G.J., De Bruin, J.P.C., Uylings, H.B.M. and Corner, M.A., eds.), Elsevier, New York, pp. 129–146.

Harlan, R.E., Shivers, B.D. & Pfaff, D.W. (1983). Midbrain microinfusions of prolactin increase the estrogen-dependent behavior, lordosis. *Science* **219**, 1451–1453.

Hasegawa, T. & Sakuma, Y. (1990). Estrogen-induced suppression of female rat forebrain neurons with axons to ventral midbrain. *Neuroscience Letters* **119**, 171–174.

Hasegawa, T. & Sakuma, Y. (1992). Sexual differentiation of oestrogen sensitivity in the rat forebrain efferent neurones. *Journal of Physiology*, in press.

Hasegawa, T., Takeo, T., Akitsu, H., Hoshina, Y. & Sakuma, Y. (1991). Interruption of the lordosis reflex of female rats by ventral midbrain stimulation. *Physiology and Behavior*, **50**, 1033–1038.

Hayes, R.L., Pechura, C.M., Katayama, Y., Povlishock, J.T., Giebel, M.L. & Becker, D.P. (1984). Activation of pontine cholinergic sites implicated in unconsciousness following cerebral concussion in the cat. *Science* **223**, 301–303.

Jenssen, C., Gruner, B., Doecke, F. & Doerner, G. (1988). Partial compensation of sexual receptivity deficits in female rats with bilateral lesions of the hypothalamic ventromedial nucleus by transplants of fetal mediobasal hypothalamic tissue. *Experimental and Clinical Endocrinology* **91**, 287–300.

Jung-Testas, I. & Hu, Z.Y. (1989) Neurosteroids: biosynthesis of pregnenolone and progesterone in primary cultures of rat glial cells. *Endocrinology* **125**, 2083–2091.

Kaitin, K.I. (1984). Preoptic area unit activity during sleep and wakefulness in the cat. *Experimental Neurology* **83**, 347–357.

Kalen, P., Skagerberg, G. & Lindvall, O. (1988). Projections from the ventral tegmental area and mesencephalic raphe to the dorsal raphe nucleus in the rat. Evidence for a minor dopaminergic component. *Experimental Brain Research* **73**, 69–77.

Ke, F.-C. & Ramirez, V.D. (1987). Membrane mechanism mediates progesterone stimulatory effect on LHRH release from superfused rat hypothalami *in vitro*. *Neuroendocrinology* **45**, 514–517.

Kelly, M.J. (1982). Electrical effects of steroids on neurons. In: *Hormonary Active Brain Peptides, Structure and Function* (McKerns, K.W. and Pantic, V., eds.), Plenum Press, New York, pp. 253–277.

Kelly, M.J., Kuhnt, U. & Wuttke, W. (1980). Hyperpolarization of hypothalamic parvocellular neurons by 17β-estradiol and their identification through intracellular staining with procion yellow. *Experimental Brain Research* **40**, 440–447.

Kendall, D.A., McEwen, B.S. & Enna, S.J. (1982). The influence of ACTH and corticosterone on [^{3}H]GABA receptor binding in rat brain. *Brain Research* **236**, 365–374.

Kendrick, D.M. & Drewett, R.F. (1980). Testosterone-sensitive neurones respond to oestradiol but not to dihydrotestosterone. *Nature* **286**, 67–68.

Kendrick, D.M., Drewett, R.F. & Wilson, C.A. (1981). Effect of testosterone on neuronal refractory periods, sexual behaviour and lutenizing hormone: a comparison of time-courses. *Journal of Endocrinology* **89**, 147–155.

King, T.R. & Nance, D.M. (1985). The effects of unilateral frontolateral hypothalamic knife cuts and asymmetrical unilateral septal lesions on lordosis behavior of rats. *Physiology and Behavior* **35**, 955–959.

Kondo, Y., Shinoda, A., Yamanouchi, K. & Arai, Y. (1986). The degree of recovery of lordotic activity by dorsal deafferentation of the preoptic area in male and androgenized female rats. *Physiology and Behavior* **37**, 495–498.

Kow, L.-M., Montgomery, M. & Pfaff, D.W. (1977). Effects of spinal cord transections on lordosis reflex in female rats. *Brain Research* **123**, 75–88.

Kow, L.-M. & Pfaff, D.W. (1976). Sensory requirements for the lordosis reflex in female rats. *Brain Research* **101**, 47–66.

Leedy, M.G. (1984). Effects of small medial preoptic lesions on estrous cycles and receptivity in female rats. *Psychoneuroendocrinology* **9**, 189–196.

Leonard, C.M. & Scott, J.W. (1971). Origin and distribution of the amygdalofugal pathways in the rat: an experimental neuroanatomical study. *Journal of Comparative Neurology* **141**, 313–330.

Lisk, R.D. & Barfield, M.A. (1975). Progesterone facilitation of sexual receptivity in rats with neural implantation of estrogen. *Neuroendocrinology* **19**, 28–35.

Lisk, R.D. & MacGregor, L. (1982). Subproestrus estrogen levels facilitate lordosis following septal or cingulate lesions. *Neuroendocrinology* **35**, 313–320.

Llinas, R. (1988). The intrinsic electrophysiological properties of mammalian neurons: Insight into central nervous system function. *Science* **242**, 1654–1664.

Lovick, T.A. (1991). Central nervous system integration of pain control and autonomic function. *News in Physiological Sciences* **6**, 82–86.

Luiten, P.G.M., Kuipers, F. & Schuitmaker, H. (1982). Organization of diencephalic and brainstem afferent projections to the lateral septum in the rat. *Neuroscience Letters* **30**, 211–216.

MacLusky, N.J. & Naftolin, F. (1981). Sexual differentiation of the central nervous system. *Science* **211**, 1294–1303.

Majewska, M.D., Harrison, N.L., Schwartz, R.D., Barker, J.L. & Paul, S.M. (1986). Steroid hormone metabolites are barbiturate-like modulators of the GABA receptor. *Science* **223**, 1004–1007.

Majewska, M.D., Mienville, J.-M. & Vicini, S. (1988). Neurosteroid pregnenolone sulfate antagonizes electrophysiological responses to GABA in neurons. *Neuroscience Letters* **90**, 279–284.

Malsbury, C.W. & Daood, J.T. (1978). Sexual receptivity: critical importance of supraoptic connections of the ventromedial hypothalamus. *Brain Research* **159**, 451–457.

Manogue, K.R., Kow, L.-M. & Pfaff, D.W. (1980). Selective brain stem transections affecting reproductive behavior of female rats: the role of hypothalamic output to the midbrain. *Hormones and Behavior* **14**, 277–302.

Marler, P. (1991). Song-learning behavior: the interface with neuroethology. *Trends in Neuroscience* **14**, 199–206.

Masco, D.H. & Carrer, H.F. (1980). Sexual receptivity in female rats after lesion or stimulation in different amygdaloid nuclei. *Physiology and Behavior* **24**, 1073–1080.

Matsumoto, A. & Arai, Y. (1986). Male-female difference in synaptic organization of the ventromedial nucleus of the hypothalamus in the rat. *Neuroendocrinology* **42**, 232–236.

McCarthy, M.M., Malik, K.F. & Feder, H.H. (1990). Increased GABAergic transmission in medial hypothalamus facilitates lordosis but has the opposite effect in preoptic area. *Brain Research* **507**, 40–44.

McEwen, B.S. (1988). Genomic regulation of sexual behavior. *Journal of Steroid Biochemistry* **30**, 179–183.

McEwen, B.S. (1991). Non-genomic and genomic effects of steroids on neural activity. *Trends in Pharmacological Science* **12**, 141–147.

McEwen, B.S., Davis, P.G., Parsons, B. & Pfaff, D.W. (1979). The brain as a target for steroid hormone action. *Annual Review of Neuroscience* **2**, 65–112.

McGinnis, M., Nance, D.M. & Gorski, R.A. (1978). Olfactory, septal and amygdala lesions alone or in combination: effects on lordosis behavior and emotionality. *Physiology and Behavior* **20**, 435–440.

Minami, T., Oomura, Y., Nabekura, J. & Fukuda, A. (1990). 17β-Estradiol depolarization of hypothalamic neurons is mediated by cyclic AMP. *Brain Research* **519**, 301–307.

Meisel, R.L. & Pfaff, D.W. (1985). Brain region specificity in estradiol effects on neuronal ultrastructure in rats. *Molecular and Cellular Endocrinology* **40**, 159–166 1985.

Mobbs, C.V., Fink, G. & Pfaff, D.W. (1990). Hip-70: a protein induced by estrogen in the brain and LH-RH in the pituitary. *Science* **247**, 1477–1479.

Mobbs, C.V., Fink, G. & Pfaff, D.W. (1990). Hip-70: an isoform of phosphoinositol-specific phospholipase C-α. *Science* **249**, 566.

Mobbs, C.V., Rothfeld, J.M., Saluja, R. & Pfaff, D.W. (1989). Phorbol esters and forskolin infused into midbrain central gray facilitate lordosis. *Pharmacology Biochemistry and Behavior* **34**, 665–667.

Morrell, J.I., Krieger, D.T. & Pfaff, D.W. (1986). Quantitative autoradiographic analysis of estradiol retention by cells in the preoptic area, hypothalamus and amygdala. *Experimental Brain Research* **62**, 343–354.

Moss, R.L., Paloutzian, R.F. & Law, O.T. (1974). Electrical stimulation of forebrain structures and its effects on copulatory as well as stimulus-bound behavior in ovariectomized hormone-primed rats. *Physiology and Behavior* **12**, 997–1004.

Nabekura, J., Oomura, Y., Minami, T., Mizuno, Y. & Fukuda, A. (1986). Mechanism of the rapid effect of 17β-estradiol on medial amygdala neurons. *Science* **233**, 226–228.

Nance, D.M., Shryne,J. & Gorski, R.A. (1975). Effects of septal lesions on behavioral sensitivity of female rats to gonadal hormones. *Hormones and Behavior* **6**, 59–64.

Nishizuka, M. & Arai, Y. (1982). Synapse formation in response to estrogen in the medial amygdala developing in the eye. *Proceedings of National Academy of Science, U.S.A.* **79**, 7024–7026.

Parsons, B., McEwen, B.S. & Pfaff, D.W. (1982). A discontinuous schedule of estradiol treatment is sufficient to activate progesterone-facilitated feminine sexual behavior and to increase cytosol receptors for progestins in the hypothalamus of the rat. *Endocrinology* **110**, 613–619.

Parsons, B., Rainbow, T.C., Snyder, L. & McEwen, B.S. (1984). Progesterone-like effects estradiol on reproductive behavior and hypothalamic progestin receptors in the female rat. *Neuroendocrinology* **39**, 25–30.

Pfaff, D.W. (1980). *Estrogens and Brain Function,* Springer, New York.

Pfaff, D.W., Lewis, C., Diakow, C. & Keiner, M. (1973). Neurophysiological analysis of mating behavior responses as hormone-sensitive reflexes. *Progress in Physiological Psychology* **5**, 253–297.

Pfaff, D.W. & Keiner, M. (1972). Atlas of estradiol-concentrating cells in the central nervous system of the female rat. *Journal of Comparative Neurology* **151**, 121–158.

Pfaff, D.W. & Sakuma, Y. (1979a). Facilitation of the lordosis reflex of female rats from the ventromedial nucleus of the hypothalamus. *Journal of Physiology* **288**, 189–202.

Pfaff, D.W. & Sakuma, Y. (1979b). Deficit in the lordosis reflex of female rats caused by lesions in the ventromedial nucleus of the hypothalamus. *Journal of Physiology* **288**, 203–210.

Pfaff, D.W. & Schwartz-Giblin, S. (1988). Cellular mechanisms of female reproductive behaviors. In: *The Physiology of Reproduction* (Knobil, E. and Neil, J., eds.), Raven Press, New York, pp. 1487–1568.

Pfaff, D.W., Silva, M.T.A. & Weiss, J.M. (1971). Telemetered recording of hormone effects on hippocampal neurons. *Science* **172**, 394–395. .

Popolow, H.B., King, J.C. & Gerall, A.A. (1981). Rostral medial preoptic area lesions' influence on female estrous processes and LHRH distribution. *Physiology & Behavior* **27**, 855–861.

Powers, B. & Valenstein, E.S. (1972). Sexual receptivity: facilitation by medial preoptic lesions in female rats. *Science* **175**, 1003–1005.

Pragnell, M., Snay, K.J., Trimmer, J.S., Maclusky, N.J., Naftolin, F., Kaczmarek, L.K. & Boyle, M.B. (1990). Estrogen induction of a small, putative K^+ channel messenger RNA in rat uterus. *Neuron* **4**, 807–812.

Quadagno, D.M., Shryne, J., Anderson, C. & Gorski, R.A. (1972). Influence of gonadal hormones on social, sexual, emergence, and open field behavior in the rat (*Rattus norvegicus*). *Animal Behavior* **29**, 732–740.

Rainbow, T.C., Snyder, L., Berck, D.J. & McEwen, B.S. (1984). Correlation of muscarinic receptor induction in the ventromedial hypothalamic nucleus with the activation of feminine sexual behavior by estradiol. *Neuroendocrinology* **39**, 476–480.

Raisman, G. & Field, P.M (1973). Sexual dimorphism in the neuropil of the preoptic area of the rat and its dependence on neonatal androgen. *Brain Research* **98**, 417–440.

Rajendren, G., Dudley, C.A. & Moss, R.L. (1990). Role of vomeronasal organ in the male induced enhancement of sexual receptivity in female rats. *Neuroendocrinology* **52**, 368–372.

Renaud, L.P. (1976). An electrophysiological study of the amygdalohypothalamic projections to the ventromedial nucleus of the rat. *Brain Research* **105**, 45–58.

Richter-Levin, G. & Segal, M. (1990). Grafting of midbrain neurons into the hippocampus restores serotonergic modulation of hippocampal activity in the rat. *Brain Research* **521**, 1–6.

Robbins, A., Schwarz-Giblin, S. & Pfaff, D.W. (1990). Ascending and descending projections to medullary reticular formation sites which activate deep lumbarback muscles in the rat. *Experimental Brain Research* **80**, 463–474.

Rodriguez-Manzo, G., Cruz, M.L. & Beyer, C. (1986). Facilitation of lordosis behavior in ovariectomized estrogen-primed rats by medial preoptic implantation of 5β, 3β-pregnenolone: a ring A-reduced progesterone metabolite. *Physiology and Behavior* **36**, 277–281.

Romano, G.J., Mobbs, C.V., Lauber, A., Howells, R.D. & Pfaff, D.W. (1990). Differential regulation of proenkephalin gene expression by estrogen in the ventromedial hypothalamus of male and female rats: implications for the molecular basis of a sexually differentiated behavior. *Brain Research* **536**, 63–68.

Sakai, K. (1980). Some anatomical and physiological properties of pontomesencephalic tegmental neurons with special reference to the PGO waves and postural atonia during paradoxical sleep in the cat. In: *The Reticular Formation Revisited* (Hobson, J.A. and Brazier, M.A.B., eds.), Raven Press, New York, pp. 427–447.

Sakuma, Y. (1984). Influences of neonatal gonadectomy or androgen exposure on the sexual differentiation of the rat ventromedial hypothalamus. *Journal of Physiology* **349**, 273–286.

Sakuma, Y. & Akaishi, T. (1987). Cell size, projection path, and localization of estrogen-sensitive neurons in the rat ventromedial hypothalamus. *Journal of Neurophysiology* **57**, 1148–1159.

Sakuma, Y. & Pfaff, D.W. (1979a). Facilitation of female reproductive behavior from mesencephalic central gray in the rat. *American Journal of Physiology* **237**, R278–R284.

Sakuma, Y. & Pfaff, D.W. (1979b). Mesencephalic mechanisms for integration of female reproductive behavior in the rat. *American Journal of Physiology* **237**, R285–R290.

Sakuma, Y. & Pfaff, D.W. (1980a). LHRH in the mesencephalic central grey can potentiate lordosis reflex of female rats. *Nature* **283**, 566–567.

Sakuma, Y. & Pfaff, D.W. (1980b). Cells of origin of medullary projections in central gray of rat mesencephalon. *Journal of Neurophysiology* **44**, 1002–1011.

Sakuma, Y. & Pfaff, D.W. (1980c). Excitability of female rat central gray cells with medullary projections: changes produced by hypothalamic stimulation and estrogen treatment. *Journal of Neurophysiology* **44**, 1012–1023.

Sakuma, Y. & Pfaff, D.W. (1982). Properties of ventromedial hypothalamic neurons with axons to midbrain central gray. *Experimental Brain Research* **46**, 292–300.

Sawyer, C.H. & Kawakami, M. (1959). Characteristics of behavioral and electroencephalographic after-reactions to copulation and vaginal stimulation in the female rabbit. *Endocrinology* **65**, 622–630.

Scalia, F. & Winans, S.S. (1975). The differential projections of the olfactory bulb and accessory olfactory bulb in mammals. *Journal of Comparative Neurology* **161**, 31–56.

Schiess, M.C., Joels, M. & Shinnick-Gallagher, P. (1988). Estrogen priming affects active membrane properties of medial amygdala neurons. *Brain Research* **440**, 380–385.

Schumacher, M. (1990). Rapid membrane effects of steroid hormones: an emerging concept in neuroendocrinology. *Trends in Neuroscience* **13**, 359–362.

Schumacher, M., Coirini, H., Pfaff, D.W. & McEwen, B.S. (1990). Behavioral effects of progesterone associated with rapid modulation of oxytocin receptors. *Science* **250**, 691–694.

Shik, M.L., Severin, F.V. & Orlovskii, G.N. (1966). Control of walking and running by means of electrical stimulation of the mid-brain. *Biofizyka* **11**, 659–666 (English translation **11**, 756–765).

Shivers, B.D., Harlan, R.E. & Pfaff, D.W. (1989). A subset of neurons containing immunoreactive prolactin is a target for estrogen regulation of gene expression in rat hypothalamus. *Neuroendocrinology* **49**, 23–27.

Simerly, R.B. & Swanson, L.W. (1988). Projections of the medial preoptic nucleus: a phaseolus vulgaris leucoagglutinin anterograde tract-tracing study in the rat. *Journal of Comparative Neurology* **270**, 209–242.

Sloviter, R.S. (1989). Calcium-binding protein (calbindin-D28k) and parvalbumin immunocytochemistry: localization in the rat hippocampus with specific reference to the selective vulnerability of hippocampal neurons to seizure activity. *Journal of Comparative Neurology* **280**, 183–196.

Sutherland, R.C., Fink, G. & Charlton, H.M. (1984). Effect of mating on metabolic activity of the brain and pituitary gland assessed by [^{14}C]2-deoxyglucose in a reflex ovulator, the vole (*Microtus agrestis*). *Brain Research* **311**, 317–322.

Swanson, L.W. (1987). The hypothalamus, In Integrated Systems of the CNS, Part I, In: *Handbook of Chemical Neuroanatomy*, Vol. 5 (Bjorklund, A. and Hokfelt, T., eds.), Elsevier, Amsterdam, pp. 1–124.

Swanson, L.W., Mogenson, J.G., Gerfen, C.R. & Robinson, P. (1984). Evidence for a projection from the lateral preoptic area and substantia innominata to the 'mesencephalic locomotor region' in the rat. *Brain Research* **295**, 161–178.

Swanson, L.W., Mogenson, J.G., Simerly, R.B. & Wu, M. (1987). Anatomical and electrophysiological evidence for a projection from the medial preoptic area to the 'mesencephalic and subthalamic locomotor regions' in the rat. *Brain Research* **405**, 108–122.

Takeo, T., Chiba, Y. & Sakuma, Y. (1992). Suppression of the lordosis reflex of female rat by efferent of the medial preoptic area. *Journal of Neuroscience*, in press.

Tong Y, Zhao, H.F., Labrie, F. & Pelletier, G. (1990). Regulation of proopiomelanocortin messenger ribonucleic acid content by sex steroids in the arcuate nucleus of the female rat brain. *Neuroscience Letters* **112**, 104–108.

Toran-Allerand, C.D. (1976). Sex steroids and the development of the newborn mouse hypothalamus and preoptic area *in vitro*: implications for sexual differentiation. *Brain Research* **106**, 407–412.

Whitney, J.F. (1986). Effect of medial preoptic lesions on sexual behavior of female rats is determined by test situation. *Behavioral Neuroscience* **10**, 230–235.

Whitten, W.K. & Champlin, A.K. (1973). The role of olfaction in mammalian reproduction. In: *Handbook of Physiology, Endocrinology*, Vol. II, Part 1 (Greep, R.O. and Astwood, E.B., eds.), American Physiological Society, Washington, D.C., pp. 109–123.

Yamanouchi, K. & Arai, Y. (1977). Possible inhibitory role of the dorsal inputs to the preoptic area and hypothalamus in regulating female sexual behavior in the female rat. *Brain Research* **127**, 296–301.

Yamanouchi, K. & Arai, Y. (1978). Lordosis behavior in male rats: effect of deafferentation in the preoptic area and hypothalamus. *Journal of Endocrinology* **76**, 381–382.

Yamanouchi, K. & Arai, Y. (1990). The septum as origin of a lordosis-inhibiting influence in female rats: effect of neural transection. *Physiology and Behavior* **48**, 351–355.

Zasorin, N.L., Malsbury, C.W. & Pfaff, D.W. (1975). Suppression of lordosis in hormone-primed female hamster by electrical stimulation of the septal area. *Physiology and Behavior* **14**, 595–599.

8 Oxytocin and Prolactin Secretion by Suckling and their Reduced Response to Stimuli Other than Suckling

Takashi HIGUCHI and Hideo NEGORO

Department of Physiology, Fukui Medical School, Fukui 910-11

Survival of all newborn mammals is dependent upon an adequate supply of milk secreted from the mammary gland of the mother. To meet a great demand for milk, mammary tissue grows rapidly and acquires the ability to secrete milk throughout pregnancy and early lactation. Among a variety of hormones controlling milk production, prolactin (PRL) has a central role in lactogenesis and maintenance of lactation. The stored milk is expelled by the action of oxytocin (OT) in contracting the myoepithelium which invests the alveoli and small ducts of the mammary gland. Both of these hormones are released in response to the somatosensory stimuli applied to the nipples by pups. Neural impulses generated by suckling are transmitted to the central nervous system, where they are integrated to result in OT release from the posterior pituitary gland and in PRL release from the anterior pituitary gland; these two mechanisms comprise typical neuroendocrine reflexes.

This chapter will present current information concerning OT and PRL release by suckling stimulation in the rat, the species that has been most well studied. Emphasis is placed on the mechanism synchronizing OT neuron burst leading to the pulsatile OT release, and on the reversible morphological and functional reorganization of the neuroendocrine systems which may facilitate pulsatile release of OT and reduced responsiveness of OT and PRL to stimuli other than suckling such as stress or estrogen.

OXYTOCIN RELEASE BY THE SUCKLING STIMULUS

Afferent Pathways

The suckling stimulus applied to the nipples by the pups excites neurons in a wide area of the dorsal horn of the spinal cord (Poulain & Wakerley, 1986). These neurons receive convergent input from several adjacent ipsilateral nipples. The ascending route after transmission through the dorsal horn has been determined by the effects of discrete lesioning the spinal cord on the milk-ejection reflex. Using rats anesthetized with urethane, we found that lesions of the lateral funiculi blocked the milk-ejection reflex, whereas lesions of the dorsal and ventral funiculus were ineffective (Fukuoka *et al.*, 1984). Ipsilateral suckling failed to elicit milk ejection in the rats with unilateral section of the lateral funiculus at the level of C6 and C7 of the spinal cord, while the milk-ejection reflex was observed in 50% of the rats during contralateral suckling. These results, together with those of Juss & Wakerley (1981) showing that milk ejection was blocked when suckling was confined to the nipples on the side contralateral to the unilaterally lesioned lateral tegmentum in the mesencephalon, suggest that the most likely route for the suckling stimulus is *via* the spinocervical tract. The spinocervical tract ascends ipsilaterally within the lateral funiculus, relaying within the lateral cervical nucleus in the cervical segments of the spinal cord (Dubois-Dauphin *et al.*, 1985).

From the mesencephalon the afferent pathway runs through the central area of the mid-hypothalamus to the supraoptic and paraventricular nuclei (Takano *et al.*, 1990a). Although a knife cut of the lateral part of the mid-hypothalamic level (5.7 mm anterior to and 0.5 mm above the interaural line, 1.3–2.8 mm from the midline) had no effect on the milk-ejection reflex, a central cut of 1 mm width on both sides abolished the reflex. Thus the major somatosensory pathway runs through the basal mid-hypothalamic area in the rat, which differs from the route taken in the guinea pig (Tindal & Knaggs, 1981). In the guinea pig the afferent pathway of the milk-ejection reflex runs within the medial forebrain bundle and maintains the lateral position until the level of the paraventricular nucleus. The reader is referred to the article by Wakerley *et al.* (1988) for more detailed information on afferent neural pathways controlling the milk-ejection reflex.

Pulsatile Release of Oxytocin

Laboratory rats nurse their young for a total of 12 to 18 hours a day, with frequent periods of nursing each lasting 20 to 30 min (Lincoln *et al.*, 1973). During the nursing period, basal OT concentrations in the blood are not different before the start of suckling from during suckling, unless the milk-ejection reflex occurs. The occurrence of the milk-ejection reflex is detected by observing stretching behavior of the pups which consists of a synchronous response from all the pups in which they pull strongly against

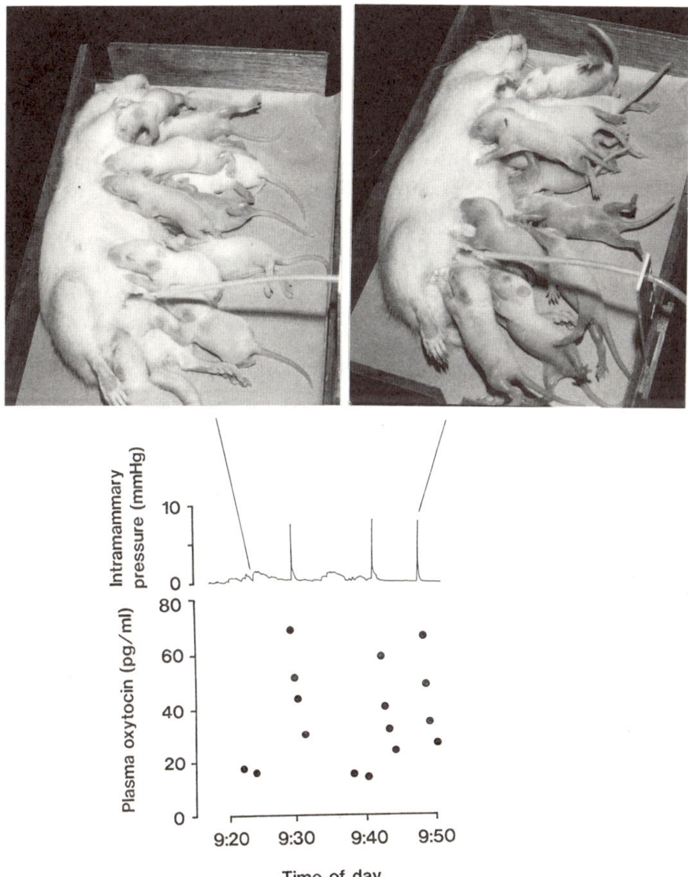

Fig. 1.   Pulsatile release of oxytocin during suckling. Plasma concentration of OT was measured in a urethane-anesthetized lactating rat by radioimmunoassay (lower panel) while monitoring intramammary pressure (upper panel). Photographs were taken at the time when the milk-ejection reflex was occurring (right) or was not occurring (left). At milk ejection, the young showed typical stretching behavior.

the nipple with their legs out stretched and their backs arched (Lincoln *et al.*, 1973). OT levels in plasma increase abruptly and uniformly by about 50 pg/ml and decline rapidly with a half-life of 1.5 min only at the milk-ejection reflex (Higuchi *et al.*, 1986). As shown in Fig.1 this characteristic pulsatile release of OT can be reproduced in rats anesthetized with urethane (Higuchi *et al.*, 1985) in which intramammary pressure can be monitored as well as the stretch behavior of the pups. This type of intermittent bolus release, of about 0.5 to 1 mU of OT, during suckling has physiological significance: this hormone message produces maximally efficient contractions of the mammary myoepithelial cells and subsequent milk let-down; as the response of the mammary gland to OT is nonlinear, doses above 1 mU do not produce proportionately larger contractions, while doses below 0.2

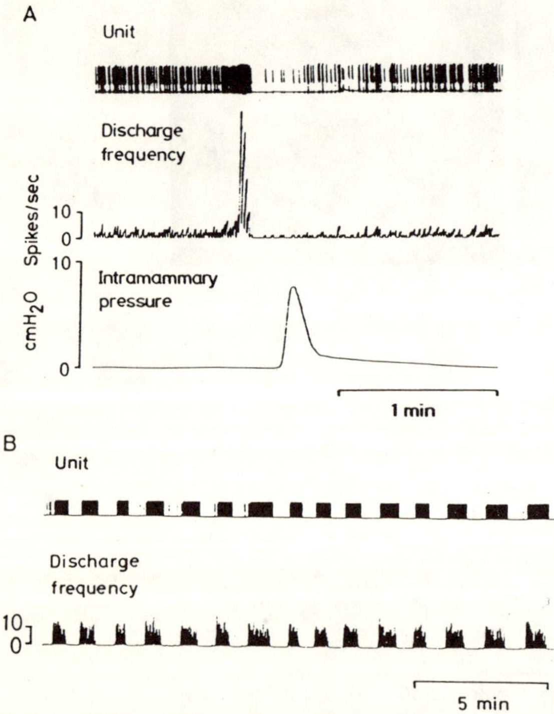

Fig. 2.   Extracellular recording of action potentials of oxytocin neuron (A) and vasopressin neuron (B) in the paraventricular nucleus in a urethane-anesthetized lactating rat. The recorded neuron is identified as extending its axon to the posterior pituitary gland by electrical stimulation of the gland. Cell types of the recorded neuron can be distinguished according to the firing pattern of bursting in the OT neuron at the milk-ejection reflex and to the phasic pattern in the vasopressin neuron.

mU have little effect. If a dose of 0.5 to 1 mU is given not as a bolus, but as an injection over 5 to 10 sec, the contraction response of the myoepithelial cells is considerably reduced (Leng, 1988). Moreover, this hormone release is achieved with a minimal number of action potentials; much more OT is released per action potential when the action potentials occur as a brief, high frequency train, than when they occur as a longer train at a lower frequency (Bicknell, 1988).

In the rat, OT-producing neurons are distributed mainly in four different cell groups, the paraventricular nucleus and the supraoptic nucleus on both sides. As shown in Fig. 2, electrical activity of these magnocellular neurons projecting to the posterior pituitary gland can be monitored by extracellular recording of their action potentials (Wakerley & Lincoln, 1973). Electrical recordings reveal that, preceding each milk ejection, probably all the oxytocin neurons show a brief burst of firing: a 20- to 40-fold acceleration for 2 to 4 sec. Apart from these intermittent high-frequency bursts of firing, the OT cells are completely refractory to the suckling stimulus. Removal and reapplication of the pups induces no apparent change in their electrical activity (Wakerley & Lincoln, 1973) or in the plasma levels of OT (Higuchi *et al.*, 1986). The intermittent pattern of OT release during suckling may provide a good experimental model for analyzing the mechanism for synchronized activation of the electrical activity of a specific group of neurons which results in pulsatile release of many kinds of hormones such as luteinizing hormone. Pulsatile release of luteinizing hormone is mentioned elsewhere in this book (Chapters 4-6).

Mechanisms for Bursting Activity in Oxytocin Neurons
A. Bursting activity in vasopressin cells
Vasopressin is the other neurohypophysial hormone in mammals, and unlike OT, is released continuously rather than episodically. Nevertheless, each vasopressin cell fires in bursts, each burst comprising 5 to 15 Hz trains of action potentials lasting 20 to 40 sec, separated by silent periods of similar duration as shown in Fig. 2. The phasically firing cells (vasopressin cells) do not fire in synchrony with each other and a neuron will secrete vasopressin during each burst, hence plasma vasopressin concentration does not fluctuate rapidly. Generally bursting can occur as an intrinsic property of an excitable cell given certain specific membrane characteristics. The bursting activity of vasopressin neurons can be reproduced in brain slice explant preparations (Hatton *et al.*, 1978). Electrophysiological studies have reveals that burst initiation is caused by Ca^{2+}-mediated, depolarizing after-potentials which summate to form plateau potentials and cause repetitive spiking (Andrew & Dudek, 1983; Bourque, 1986). Burst

termination involves an after-hyperpolarization caused by a Ca^{2+}-activated K^+ conductance (Andrew & Dudek, 1984; Bourque *et al.*, 1985). In addition to the intrinsic properties of the magnocellular neuroendocrine cells, the bursting pattern can be modulated by synaptically-mediated influences on the membrane and by second messengers within the bursting cells (Randle *et al.*, 1986).

B. Bursting activity in oxytocin cells

In contrast to the burst activity of vasopressin neurons, that of OT neurons is not reproducible *in vitro*, hence the bursting mechanism of OT neurons is more difficult to analyze. Although intensive experiments to study the membrane properties of identified OT cells involved in milk-ejection bursts have not been performed, it is likely that intrinsic membrane properties similar to those of vasopressin neurons may also be involved in the burst firing of OT neurons. The intrinsic characteristic of the OT neuron itself, however, cannot explain the synchronizing bursting activity at the milk-ejection reflex in almost all OT neurons which are separately located in the four nuclei in the hypothalamus.

There are several hypotheses for the mechanism of the synchronization (Leng, 1988). In the supraoptic nuclei of non-lactating rats, glial cells have processes which separate neurosecretory cells from each other, and less than 10% of these cells are involved in direct soma-soma contacts or soma-dendritic contacts. Interestingly, lactating rats show a reversible dramatic structural reorganization of the supraoptic nuclei. From late pregnancy the processes of glial cells begin to retract, leaving more cells in direct contact with other cells. In lactating rats more than 30% of the cells were involved in direct cell to cell contacts (Theodosis & Poulain, 1984), which may provide a mechanism for nonsynaptic mutual excitation either by field effects or by changes in extracellular potassium concentration (Leng & Shibuki, 1987). There is also some evidence that the number of gap junctions between OT cells is increased in lactating rats, enabling direct electrical coupling between OT neurons (Cobbett *et al.*, 1986). Moreover, there is evidence that the number of "double" synapses, which make synaptic contacts with each of two adjacent supraoptic neurons, is greatly increased in the lactating rat (Theodosis & Poulain, 1984). These morphological changes may provide a coupling of OT cells with each other during suckling in lactating rats.

There is some morphological and electrophysiological evidence for the existence of recurrent neuronal networks within the paraventricular nucleus and the supraoptic nucleus. Many synaptic endings on magnocellular neurons of the supraoptic nucleus (Leranth *et al.*, 1975) or the paraventricular nucleus (Kiss *et al.*, 1983) come from neurons in or close to the magnocellular nuclei. There is also electrophysiological evidence for functional neural connections between the ipsilateral supraoptic and paraventricular nuclei (Saphier & Feldman, 1985) and between the two supraoptic nuclei on both ipsi- and contralateral sides (Takano *et al.*, 1990b). These findings may support the hypothesis that OT neurons are coupled not only within one magnocellular nucleus but also between the nuclei directly or indirectly *via* interneurons in a positive feedback network. However, electrophysiological study to confirm such coupling is not forthcoming (Dyball & Leng, 1986). Simulating of the initial part of the milk-ejection burst was attempted by applying short trains of electrical stimulation to the neural stalk to excite OT neurons, but this failed to trigger milk-ejection bursts (Dyball & Leng, 1986). Also, high-frequency stimulation of the supraoptic nucleus which mimics OT cell burst activity at the milk ejection failed to activate contralateral OT cells (Takano *et al.*, 1990b). Thus a simple positive feedback circuit between four magnocellular nuclei may not explain OT neuron burst activity during suckling.

Afferent stimuli leading to milk ejection may be in a bursting form further back in the pathway to OT cells. In this context, we have found interesting new types of neurons in the medial part of the dorsomedial and posterior hypothalamic nuclei. These neurons do not send their axons to the posterior pituitary but to the supraoptic nucleus, and they display a brief high-frequency burst of spikes before milk ejection like the OT neuron itself (Takano *et al.*, 1991). We do not know whether these neurons play the role of "gate" or pulse generator or relay the already synchronized input to the OT neurons. However, since they exist in a region whose destruction abolishes the milk-ejection reflex (Yokoyama & Ota, 1959; Takano *et al.*, 1990a), these neurons are likely to play an essential role in the milk-ejection burst activity of OT neurons.

C. *Oxytocin role in facilitating oxytocin neuron bursts*

Freund-Mercier and Richard reported that injection of small amounts of OT (less than 1 ng) into the lateral or third ventricle induced a facilitation of the milk-ejection reflex; OT enhanced cell recruitment, periodicity and amplitude of bursts, whereas OT antagonist had the opposite effects (Freund-Mercier & Richard, 1984). They studied local release of OT in the supraoptic nucleus by push-pull perfusion technique and found that OT was released locally within the supraoptic nucleus in response to the suckling stimulus (Moos *et al.*, 1989). OT may be released from presynaptic terminals (Theodosis, 1985) and/or the extensive dendritic field of the OT neurons. Pow and Morris (1989) have visualized exocytosis of neurosecretory granules from dendrites of magnocellular neurons by using tannic acid, which stabilizes extracellular peptidergic granule cores. Furthermore, there is a report that the release of OT and vasopressin is not affected by tetrodotoxin (Di Scala-Guenot *et al.*, 1987), indicating dendritic release analogous to the reported release of dopamine from dendrites in the nigrostriatal nucleus (Greenfield, 1985).

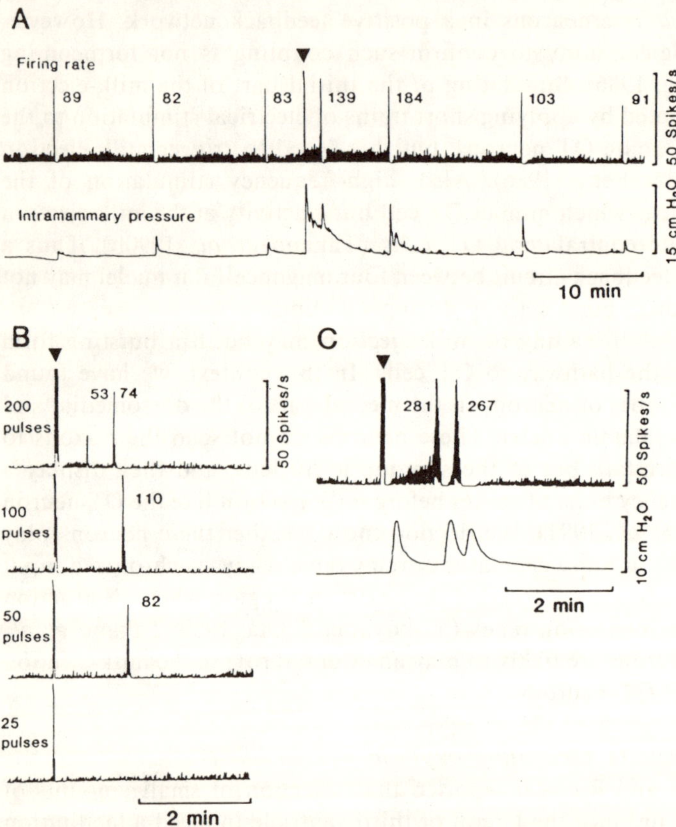

Fig. 3. Facilitation of milk-ejection reflex by electrical stimulation of the posterior pituitary gland. A: Stimulation of the neurohypophysis (200 pulses; 1 mA, 0.5-ms duration pulse for 4 s) as indicated by solid triangles, induced an additional burst of firing of an OT neuron in a urethane-anesthetized lactating rat which, except for the additional one induced by the stimulation, was exhibiting regular milk ejections. The total number of action potentials per burst episode is shown. B: Increasing the stimulation intensity evoked a burst with higher amplitude and shorter latency. C: Electrical stimulation of the neurohypophysis increased the firing rate of background activity of OT cells as well as inducing additional bursts (From Negoro *et al.*, 1985).

The released OT in the paraventricular nucleus can act directly on the neurons in the same nucleus to increase their action potentials; OT excited 57% of paraventricular nucleus cells and inhibited 3% at a dose of 10^{-7} M in the rat hypothalamic slice preparation, and this facilitatory effect was maintained after blocking synaptic transmission with low Ca^{2+} and high Mg^{2+} medium (Inenaga & Yamashita, 1986). Injecting OT into a magnocellular nucleus facilitated the occurrence and increased the amplitude of bursts of the OT cells in both the contralateral paraventricular and supraoptic nuclei (Moos & Richard, 1989). Thus OT released in the extracellular fluid of the magnocellular nuclei may render OT cells capable of bursting. However, autoradiographic studies of OT binding in the brain have failed to demonstrate high affinity binding sites in the paraventricular or supraoptic nucleus (Tribollet *et al.*, 1988). The supraoptic nucleus is probably not the site of action of intracerebroventricularly injected OT because it is located too far from the ventricle. Although the paraventricular nucleus is located at a more accessible site for OT injected into the cerebral ventricle, lesioning the paraventricular nucleus did not entirely block the facilitatory effect of OT injected intracerebroventricularly on the milk-ejection reflex (Wakerley *et al.*, 1990).

Electrical stimulation of the neurohypophysis to excite OT cells antidromically fails to trigger milk-ejection burst (Dyball & Leng, 1986), but it does facilitate the milk-ejection reflex (Fig.3; Negoro *et al.*, 1985; Dyball & Leng, 1986). Stimulation of the neurohypophysis at 50–300 pulse/s for 0.5–6 s with 1 mA evoked additional, higher amplitude bursts (Negoro *et al.*, 1985). This finding may suggest that endogenous OT released within the magnocellular nuclei affects the OT neurons to increase their ability to burst, but it is also possible that electrical stimulation may excite neurons within the nuclei projecting to the gating center such as the bed nucleus of the stria terminalis (Moos *et al.*, 1991) which shows very high specific binding of OT (Tribollet *et al.*, 1985), and can modify the periodicity of the OT cell bursts (Moos *et al.*, 1991). The evidence available at present does not allow clear establishment of how and where endogenous OT is involved in bursting activity in OT neuron at the milk-ejection reflex.

PROLACTIN RELEASE BY THE SUCKLING STIMULUS

Afferent Pathways

Few studies have attempted to trace the afferent pathway for PRL release from the nipple to the midbrain, but it is probably true that the signal for this release may share a common route with that for OT release up to a certain level of the central nervous system. Tindal and Knaggs (1977) demonstrated the ascending pathway for PRL release by measuring PRL after electrical stimulation of discrete sites in the forebrain of lactating rats. The animals they used were decorticated except for the frontal cortex, and their brainstems were transected at the mid-collicular level to avoid the influence of anesthesia. They demonstrated that the ascending pathway, probably involving the dorsal longitudinal fasciculus, enters the caudal hypothalamus from the midbrain on a broad front. It passes rostrally in the dorsal periventricular area, moves laterally to the mediodorsal and then to the lateral hypothalamus, and continues rostrally to the lateral preoptic area within the medial forebrain bundle. From the preoptic area the path swings in from the lateral to the medial preoptic area and then turns caudally for a short distance before terminating in the rostral periventricular part of the anterior hypothalamic area. This rostral periventricular region in the anterior hypothalamic area projects monosynaptically to the arcuate nucleus (Kawakami & Sakuma, 1976), and is an effective site for stimulating PRL release in the rat (Kawakami *et al.*, 1973). Since the pathway described above is based on evidence of effective sites for electrical stimulation of PRL release, the path is not necessarily involved in suckling-induced PRL release but may be involved in other PRL releasing stimuli. More detailed analysis of the afferent pathways for suckling-induced PRL secretion compared with that for OT secretion are needed to give greater insight into the mechanisms underlying the different patterns of release of PRL and OT during suckling.

Prolactin Release Inhibiting Factor and Prolactin Releasing Factor
A.   Prolactin release inhibiting factor
It is well established that the hypothalamus exerts a tonic inhibitory influence upon PRL secretion since PRL secretion occurs at a higher rate when the pituitary gland is transplanted to a site distant from the hypothalamus. Dopamine is a physiologically important prolactin release inhibiting factor (PIF) which is secreted into hypothalamic portal blood to inhibit tonically PRL release from lactotropes (MacLeod, 1976). The principal evidence supporting this view is that dopamine receptors exist on lactotropes (Brown *et al.*, 1976), and that dopamine concentrations in hypophysial stalk plasma are sufficient to inhibit PRL release (Ben-Jonathan *et al.*, 1977). There are several convincing reports of the existence of a hypothalamic PIF other than dopamine, such as γ-aminobutyric acid (Schally *et al.*, 1977) and gonadotropin-releasing hormone associated peptide (Nikolics *et al.*, 1985). However, the physiological role of these factors in controlling PRL release remains to be determined.

B. *Prolactin releasing factor*

The main stimuli that elevate PRL secretion above baseline are suckling, stress and the ovarian hormones in the rat. The increase in PRL secretion following these stimuli may be induced by hypothalamic secretion of a PRL releasing factor (PRF) rather than a decrease in dopamine secretion into the hypophysial portal vessels. There are a lot of naturally occurring compounds that will release PRL by a direct action on the pituitary gland, such as thyrotropin releasing hormone (TRH), vasoactive intestinal peptide (VIP), OT, β-endorphin, enkephalins, neurotensin, substance P, vasopressin, bombesin, angiotensin II, gonadotropin releasing hormone and others (Neill, 1988). The PRL-releasing action of these compounds probably includes a paracrine or an autocrine agent which mediates intercellular communication or acts as a mediator to generate another paracrine agent which actually affects PRL release (Jones *et al.*, 1990). These peptides have not been examined in sufficient detail to be considered strong candidates as physiological PRF, except for TRH, VIP and OT for which there is considerable evidence for PRF action (Neill, 1988). These hormones are found in portal blood at higher concentrations than in peripheral plasma, suggesting that they are released from the median eminence. Passive immunization employing specific antibodies generated against each peptide suppresses PRL secretion seen during conditions of physiological PRL release (Neill, 1988).

In addition, Ben-Jonathan and co-workers recently reported that posterior pituitary lobectomy blocked suckling-induced PRL release (Murai & Ben-Jonathan, 1987). They further demonstrated that the posterior pituitary PRF was not TRH, VIP or oxytocin by using anterior pituitary cell cultures and specific antagonists for each of the peptides (Hyde *et al.*, 1987). However, one report indicates that OT is the major PRF in the posterior pituitary gland (Mori *et al.*, 1990). Neill and co-workers reported that the vasopressin related glycopeptide was capable of stimulating PRL release (Nagy *et al.*, 1988), but the posterior pituitary glands from the Brattleboro rat, which lacks vasopressin and its associated glycoprotein, contain equivalent amounts of PRF as do other strains (Hyde *et al.*, 1989). Thus the chemical structure of the postulated posterior pituitary PRF is unknown at present.

Hypothalamic Factors Involved in Suckling-induced Prolactin Release

A. *Prolactin release by suckling*

When mother rats are separated from their young for several hours, the initiation of suckling by reunited pups is followed by a rapid increase in

PRL concentrations in the general circulation which first becomes significant within 5 min (Terkel *et al.*, 1972; Higuchi *et al.*, 1983) as seen in Fig. 4. Blood PRL concentrations exhibit pulsatile fluctuations in spite of the continuous suckling by the pups, with peaks at 6–15 min intervals. This pulsatility of PRL release was confirmed in urethane-anesthetized lactating rats whose pups could continuously suck without interference by the mother's behavior (Fig. 4; Higuchi *et al.*, 1983). Separation of lactating rats from their litters results in a marked and complete blockade of PRL release as in the case of hypophysectomy, suggesting rapid reactivation of

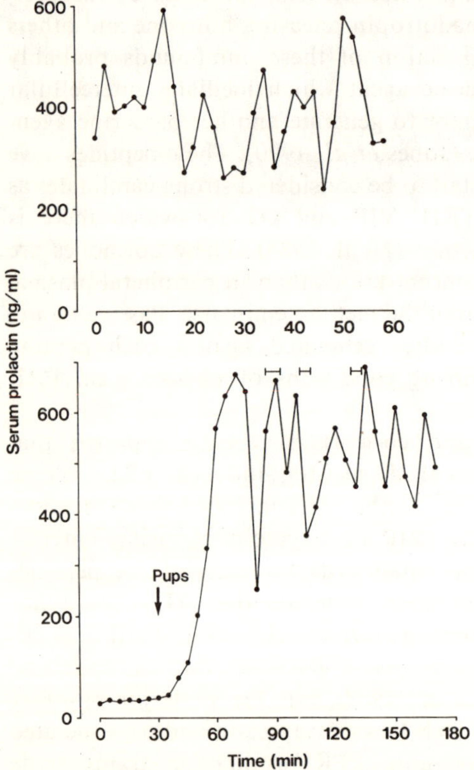

Fig. 4.   Prolactin release by suckling stimulus. Two examples of the secretory patterns of serum PRL during nursing in a conscious (lower panel) and urethane-anesthetized rat (upper panel). Pups separated from the mother for 10 h then reunited at the time shown by the arrow. Horizontal lines indicate the period when the pup did not suckle. Serum PRL levels also revealed a pulsatile pattern in the urethane-anesthetized rat, in which the young were able to suckle without being interrupted by the mother's behavior. Blood samples were taken at 5-min (lower panel) or 2-min (upper panel) intervals. (From Higuchi *et al.*, 1983).

dopaminergic inhibition and/or immediate blockade of PRF secretion (Nagy & Halász, 1983). Which inhibiting and/or releasing factor(s) are involved in suckling-induced PRL release? The simplest explanation for the PRL release is that the suckling stimulus reduces the tonic inhibition by the hypothalamus, freeing the pituitary gland to secrete PRL at a high rate.

B.   Tuberoinfundibular dopaminergic neuronal activity during suckling
Dopaminergic perikarya in the arcuate nucleus, A12 region according to the designation by Dahlström & Fuxe (1964), project to the median eminence. These neurons constitute the tuberoinfundibular system whose activity determines the amount of dopamine secreted into the hypophysial portal blood. A number of experiments have estimated the activity of the tuberoinfundibular dopaminergic neurons during suckling. The rate of decline of dopamine following administration of α-methyl-p-tyrosine (tyrosine hydroxylase inhibitor), measured by using a histofluorescence technique, was reported to be enhanced during lactation (Fuxe *et al.*, 1969). However more recent studies using biochemical measurement of the accumulation of L-3,4-dihydroxyphenylalanine (L-DOPA) in the median eminence following injection of DOPA decarboxylase inhibitor, have demonstrated a decrease in the activity of tuberoinfundibular dopaminergic neurons at the time when serum PRL was high in response to suckling (Demarest *et al.*, 1983). Thus, these results seem to support the hypothesis that suckling inhibits tonic dopaminergic inhibition, resulting in a disinhibition of PRL secretion. The dopamine concentration in hypophysial portal blood has been measured in urethane-anesthetized lactating rats in which suckling was simulated by the electrical stimulation of an isolated mammary nerve trunk. Dopamine levels in hypophysial stalk plasma decreased by 20% during electrical stimulation for 15 min, returned to prestimulation values, and then increased significantly by 20% at 45-60 min compared with unstimulated animals (de Greef *et al.*, 1981). More detailed analysis of the pattern of dopamine secretion resulting from the simulated suckling was performed with the use of electrochemical probes for measurement of dopamine release *in situ* in the median eminence (Plotsky & Neill, 1982a). The results revealed a brief but profound decrease (60-70%, lasting only for 3-5 min) in dopamine concentration immediately after the onset of electrical stimulation of a mammary nerve trunk (Plotsky & Neill, 1982a). However, simulation of the decrease in dopamine release by infusion of this drug in the physiological range to the rat in which endogenous dopamine secretion was blocked by α-methyltyrosine, demonstrated that such a small decrease of dopamine could not account for the large increase in PRL release during suckling (Plotsky & Neill, 1982b).

In the experiments to measure portal blood dopamine, the animals were in an extremely unphysiological condition (under anesthesia; surgically prepared for collection of hypophysial stalk blood; and suckling stimulus was mimicked by electrical stimulation of one isolated mammary nerve trunk). Push-pull perfusion of the median eminence-arcuate nucleus area could avoid these disadvantages and have demonstrated that suckling induces a 50% decrease in the release of dopamine which lasts for 15–30 min during the 90-min suckling period (Fig. 5, Rondeel *et al.*, 1988). From these results, it may be concluded that tuberoinfundibular dopaminergic neuronal activity is reduced by suckling stimulus, but a simple, negative, mirror image relationship between dopamine levels in the hypophysial portal blood and prolactin secretion does not exist. It remains to be elucidated why disinhibition of dopaminergic neurons by a real or a simulated suckling stimulus does not last for the entire period of stimulation: desensitization, depletion of releasable dopamine, adaptation or some other mechanism? Whatever the mechanism of short-lasting dopaminergic

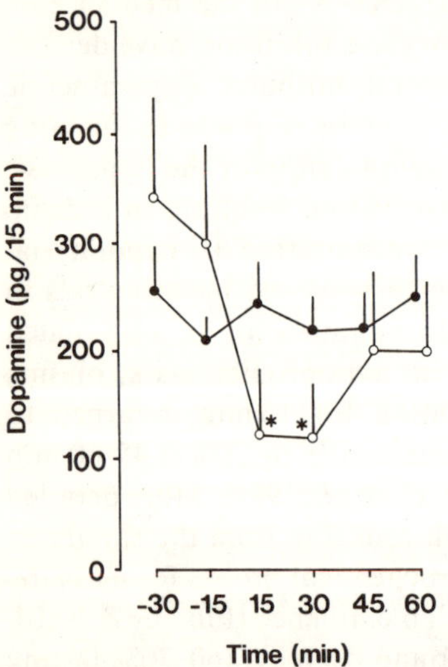

Fig. 5.   Dopamine release by suckling stimulus. Dopamine release into the median eminence-arcuate nucleus area was determined using push-pull perfusion in conscious mother rats during suckling on day 8 of lactation. Perfusate collected for 15 min was assayed for dopamine. Mother rats were allowed (○) or not to (●) nurse their pups following a 6 h separation. Data are mean ± SEM of 9–10 animals. Suckling significantly ($P < 0.05$) reduced dopamine levels for 30 min after the onset of suckling (Rondeel *et al.*, 1988). * Values significantly different from those before introduction of pups (time −30 to 0 min).

disinhibition, only a brief disinhibition may be enough to sensitize or prime the pituitary gland to subsequent PRF stimuli (Grosvenor & Mena, 1982). This sensitization process is manifest in the lactotropes as transformation of PRL to a less soluble and releasable form (Grosvenor & Mena, 1982), or at the cellular level as proportional shifts towards the cells most responsive to stimulatory secretagogues and away from those most susceptible to inhibition by dopamine (Nagy & Frawley, 1990).

C. Prolactin releasing factor involvement in suckling-induced prolactin release

There must be an involvement of some PRF(s) in suckling-induced PRL release as well as an inhibition of DA secretion since the changes in dopamine levels in the hypophysial portal blood cannot fully explain the PRL increase during suckling. Any PRF candidate listed in the section on PIF and PRF may be a physiological PRF mediating the stimulation of PRL secretion by suckling, but TRH, VIP and OT are relatively well studied with respect to their roles as PRFs, including in response to suckling. Passive immunization with antiserum against TRH (de Greef *et al.*, 1987), VIP (Abe *et al.*, 1985), or OT (Samson *et al.*, 1986) inhibits PRL secretion induced by suckling. The effect of each antiserum is not complete, suggesting that no single PRF of these candidates accounts for the total PRL release during suckling, and that a complex interaction of PRFs in addition to DA inhibition is necessary to induce PRL secretion in response to suckling. Further studies on these putative PRF candidates and other possible PRFs mentioned earlier must be performed to establish PRF(s) that physiologically mediate suckling-induced PRL release.

Surgical removal of the posterior pituitary gland completely prevents the PRL increase during suckling (Murai & Ben-Jonathan, 1987). Therefore in contrast to other putative PRFs, the posterior pituitary PRF(s) appear to be essential for suckling induced PRL release. Isolation of the posterior pituitary PRF may solve the complex problem concerning the involvement of PRF in PRL secretion in response to suckling.

REDUCED RESPONSE IN LACTATION OF OXYTOCIN AND PROLACTIN SECRETION TO STIMULI OTHER THAN SUCKLING

Stimuli Affecting Oxytocin Release Besides Suckling

Secretion of OT increases during parturition to facilitate the expulsion of fetuses and at the milk ejection to expel milk (Higuchi *et al.*, 1986). In addition to these well known physiological stimuli, in the rat OT secretion increases by hyperosmotic stimulation (Brimble *et al.*, 1978; Negoro *et al.*,

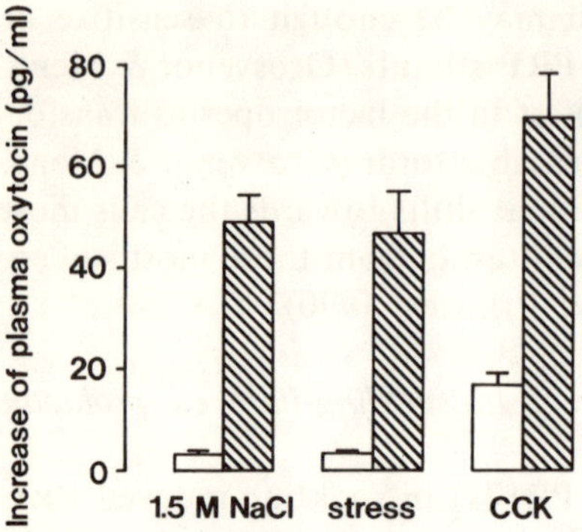

Fig. 6.   Reduced oxytocin secretory response in lactating rats. OT release in response to osmotic stimulation (intravenous injection of 1.2 ml/kg of 1.5 mol/*l* saline), immobilization stress and cholecystokinin 1–8 (intravenous injection of 20 μg/kg) was examined in lactating (□) and in non-lactating (▨) rats. OT increase from prestimulation levels at 5 min, 30 min and 10 min after the stimulation is shown, respectively (Higuchi *et al.*, 1988; Higuchi *et al.*, 1991a). Data are mean±SEM of 6–8 animals.

1988), various kinds of stress (Lang *et al.*, 1983), nausea (Verbalis *et al.*, 1986b) and food intake (Verbalis *et al.*, 1986a), although the physiological meaning of the increased OT secretion in these situations is not well understood. As shown in Fig.6, lactating rats exhibit reduced OT release compared with virgin females in response to stress (Carter & Lightman, 1987; Higuchi *et al.*, 1988), to osmotic stimulation (Higuchi *et al.*, 1988) and to cholecystokinin which may be related to nausea (Higuchi *et al.*, 1991a).

To seek the origin of this reduced sensitivity we compared the electrical responsiveness of OT neurons as well as their OT releasing ability, and extracellular potassium concentrations of isolated neural lobes between lactating and virgin female rats (Higuchi *et al.*, 1991a). A clear reduced OT release was observed in lactating rats under urethane anesthesia in response to systemic administration of cholecystokinin, which is a strong OT secretogogue probably associated with activation of central nausea pathways mediated by the vagus nerve (Verbalis *et al.*, 1986a). Electrophysiological recordings of single cells in the supraoptic nucleus, however, showed no difference in the responsiveness of OT cells to cholecystokinin. The extracellular potassium response to electrical stimulation of the isolated neural lobe, as an index of the excitability of neural lobe neurosecretory axons, was unchanged in lactating compared with virgin female rats in spite of reduced OT release (Higuchi *et al.*, 1991a). These results indicate that the reduced OT release in lactating rats is a consequence of reduced OT content in the neural lobe rather than of a reduced excitability of the oxytocin neurons either at the soma or axon terminal level (Higuchi *et al.*, 1991a).

This result, nevertheless, does not exclude the possibilities that there is reduced responsiveness at the level of OT neurons of lactating rats to some neurotransmitter(s) which transmits certain specific stimuli and at the level of the previous steps to OT neurons. In fact, supraoptic cells from hypothalamic slices of lactating rats show reduced sensitivity to α-adrenergic receptor excitation (Wakerley & Negoro, 1990).

Reduced Response to Stress

Intracerebroventricular administration of corticotropin-releasing factor (CRF) produced a dose-dependent rise in the plasma OT concentration (Bruhn *et al.*, 1987; Higuchi *et al.*, 1990). Lesioning of the paraventricular nucleus (which contains the majority of CRF neurons), reduces OT release in response to immobilization stress. Anti-CRF serum injection into the third ventricle also reduces delayed part of OT response to immobilization stress (Higuchi *et al.*, 1990). Thus, OT release seen during stressful conditions could be added to the general stress responses which are supposed to be mediated by CRF (Lenz *et al.*, 1987), although the precise mechanism by which CRF affects OT release is unknown.

It has been reported by Patel *et al.* (1991) that in lactating rats there is a complete abolition of the OT release induced by intracerebrospinal injection of CRF. Thus, the reduced OT response to stress in lactating rats (Carter & Lightman, 1987; Higuchi *et al.*, 1988) could be due to lower releasable OT content in the posterior pituitary as described above (Higuchi *et al.*, 1991a), to the reduced sensitivity to α-adrenergic stimulation (Wakerley & Negoro, 1990) and/or due to low responsiveness of OT neurons to endogenously released CRF as suggested by Patel *et al.* (1991). Moreover endogenous CRF release affecting OT neurons as well as CRF secretion into the hypophysial portal vessels may be blunted in lactating rats, as suggested by the finding that lactation interrupts the stress-mediated activation of CRF messenger RNA accumulation, while leaving intact the adrenalectomy-mediated increase in CRF mRNA (Lightman & Young, 1989).

A major feature of lactation is the marked blunting of the hormonal responses to stress, which has been described for ACTH/corticosterone (Thoman *et al.*, 1970; Stern *et al.*, 1973), PRL (Fig. 7; Stern & Voogt, 1973/74; Carter & Lightman, 1987; Higuchi *et al.*, 1988) and catecholamine (Higuchi *et al.*, 1989). However, lactation is associated with major changes in maternal aggressive behavior (Thoman *et al.*, 1970; Erskine *et al.*, 1978), water balance (Kaufman, 1981), glucose metabolism (Burnol *et al.*, 1986) as

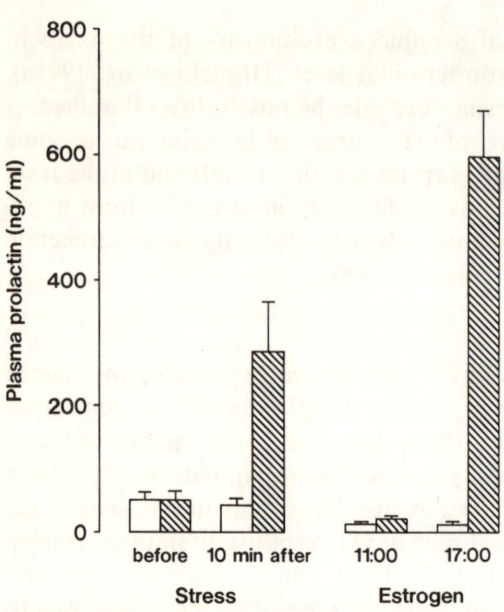

Fig. 7.   Reduced prolactin secretory response in lactating rats. Changes in serum PRL after 10 min immobilization stress and estrogen treatment were examined in lactating (□) and non-lactating rats (▨). For details of the estrogen experiment, see text. Data are mean±SEM of 8-13 animals. (From Higuchi *et al.*, 1989; Higuchi *et al.*, 1992a).

well as neuroendocrine function. Thus, reduced OT and PRL responsiveness to stress could be one aspect of the considerable adaptive changes in the emotional and autonomic functions of lactating animals.

Reduced Response to Estrogen

A large amount of PRL is secreted during the afternoon of proestrus in female rats. This PRL surge can be demonstrated by estradiol benzoate administration into ovariectomized rats (Caligaris *et al.*, 1974; Higuchi *et al.*, 1992a); 2 days after the estrogen treatment serum PRL levels increase in the afternoon with a peak at 17:00 h. However, the same estrogen treatment on days 8-10 postpartum fails to induce a PRL surge in ovariectomized lactating rats whose pups are removed at 9:00 h 2 days after estrogen injection (Fig.7). Since there is no difference in PRL response to TRH and dopamine antagonist given just before the expected PRL surge, the failure to induce a PRL surge in lactating rats may not be attributable to the anterior pituitary level but to the functional changes in the central nervous system. The suckling stimulus may alter the sensitivity to estrogen stimulation in the neurons which control PRL release, leading to an inhibition of the PRL surge.

Our recent results on tuberoinfundibular dopaminergic neuronal activity obtained by measuring the DA metabolite, 3,4-dihydroxyphenylacetic acid in the median eminence, have shown that tuberoinfundibular dopaminergic neuronal activity is low at the time of the PRL surge induced by estrogen administration in the ovariectomized non-lactating rat; this indicates inhibition of dopaminergic influence upon PRL release but this inhibition of the dopaminergic neurons did not occur in the ovariectomized lactating rat (Higuchi *et al.*, 1992b). The finding suggests that the tuberoinfundibular dopaminergic neuron is a likely candidate which exhibits functional changes in the regulation of PRL, at least for estrogen-mediated PRL release in lactating rats. It remains to be determined whether a functional change of the tuberoinfundibular dopaminergic neuron itself is responsible for reduced PRL release, or is a reflection of altered function of other area(s) controlling tuberoinfundibular dopaminergic neuronal activity in the lactating rat, and whether suppression of dopaminergic inhibition is the mechanism for reduced PRL responsiveness to stimuli other than estrogen, such as stress.

Lactation causes major structural and functional changes in the hypothalamo-pituitary system which favor provision of milk to pups during suckling and shut off OT and PRL secretory responses to other stimuli that are very effective in causing OT and PRL release in non-lactating animals. There appears to be a "switch" mechanism which operates during the lactating period, allowing the suckling stimulus, but not others, to activate OT and PRL secretion. Elucidating the mechanism for the reduced response of OT and PRL to stimuli other than suckling may deepen understanding of the response of hormones to the suckling stimulus itself.

CONCLUSION

We have reviewed the studies on OT and PRL secretion in response to the suckling stimulus and some other stimuli. OT release at milk ejection is the most well understood neuroendocrine reflex; it is characterized at the identified single OT neuron level. However, there remain many questions to be elucidated, among which the mechanism for synchronizing burst discharge of OT cells distributing in four discrete hypothalamic nuclei during suckling is particularly fascinating. The discovery of facilitation of the synchronizing burst by OT itself and of the neurons which display synchronizing burst on the afferent pathway to the OT cells may provide important clues to study the function of the 'gate' which transforms the continuous suckling stimulus into intermittent synchronized burst discharges of OT neurons. For PRL release during suckling, it is necessary to specify exactly the PRF(s) involved in the response, although it is generally accepted that tuberoinfundibular dopaminergic neuronal activities are decreased but that the reduction is not enough to account for the PRL release during the nursing period. It seems promising to isolate PRF in the posterior pituitary gland and to examine the effects of PRF candidates as paracrine or autocrine factors in regulating PRL secretion. Recent findings on the reduced OT and PRL response to stimuli other than suckling in the lactating rat suggest that suckling may induce functional alterations in the mechanisms controlling OT and PRL secretion, which enables the rat to adapt to the large demands for these hormones during the lactating period. Clarifying the mechanism for the reduced response of OT and PRL may be important for understanding of the hormone response to the suckling stimulus itself.

ACKNOWLEDGEMENT

We are grateful to Dr. G. Leng (Institute of Animal Physiology and Genetics Research, Cambridge), and Dr. J.A. Russell (University of Edinburgh) for critical reading of the manuscript and to Mrs. R. Yamada for excellent secretarial assistance.

REFERENCES

Abe, H., Engler, D., Molitch, M.E., Bollinger-Gruber, J. & Reichlin, S. (1985). Vasoactive intestinal peptide is a physiological mediator of prolactin release in the rat. *Endocrinology* **116**, 1383–1390.

Andrew, R.D. & Dudek, F.E. (1983). Burst discharge in mammalian neuroendocrine cells involves an intrinsic regenerative mechanism. *Science* **221**, 1050–1052.

Andrew, R.D. & Dudek, F.E. (1984). Intrinsic inhibition in magnocellular neuroendocrine cells of rat hypothalamic slices. *Journal of Physiology* **353**, 171–185.

Ben-Jonathan, N., Oliver, C., Weiner, H.J., Mical, R.S. & Porter, J.C. (1977). Dopamine in hypophysial portal plasma of the rat during the estrous cycle and throughout pregnancy. *Endocrinology* **100**, 452–458.

Bicknell, R.J. (1988). Downstream consequences of bursting activity in oxytocin neurones. In: *Pulsatility in Neuroendocrine Systems* (Leng, G., ed.), CRC Press, Boca Raton, pp. 61–74.

Bourque, C.E. (1986). Calcium-dependent spike after-current induces burst firing in magnocellular neurosecretory cells. *Neuroscience Letters* **70**, 204–209.

Bourque, C.W., Randle, J.C.R. & Renaud, L.P. (1985). A calcium-dependent potassium conductance in rat supraoptic nucleus neurosecretory neurons. *Journal of Neurophysiology* **54**, 1375–1382.

Brimble, M.J., Dyball, R.E.J. & Forslig, M.L. (1978). Oxytocin release following osmotic activation of oxytocin neurones in the paraventricular and supraoptic nuclei.

Journal of Physiology **278**, 69-78.

Brown, G.M., Seeman, P & Lee, T. (1976). Dopamine/neuroleptic receptors in basal hypothalamus and pituitary. *Endocrinology* **99**, 1407-1410.

Bruhn, T.O., Sutton, .W., Plotsky, P.M. & Vale, W.W. (1987). Central administration of corticotropin-releasing factor modulates oxytocin secretion in the rat. *Endocrinology* **119**, 1558-1563.

Burnol, A.-F., Leturque, A., Ferré, Kande, J. & Girard, J. (1986). Increased insulin sensitivity and responsiveness during lactation in rats. *American Journal of Physiology* **251**, E537-E541.

Caligaris, L., Astrada, J.J. & Taleisnik, S. (1974). Oestrogen and progesterone influence on the release of prolactin in ovariectomized rats. *Journal of Endocrinology* **60**, 205-215.

Carter, D.A. & Lightman, S.L. (1987). Oxytocin responses to stress in lactating and hyperprolactinaemic rats. *Neuroendocrinology* **46**, 532-537.

Cobbett, P., Smithson, K.G. & Hatton, G.I. (1986). Dye-coupled magnocellular peptidergic neurones of the rat paraventricular nucleus show homotypic immunoreactivity. *Neuroscience* **16**, 885-895.

Dahlström, A. & Fuxe, K. (1964). Evidence for the existence of monoamine-containing neurons in the central nervous system. *Acta Physiologica Scandinavica (Supplement)* **62**, 1-52.

de Greef, W.J., Plotsky, P.M. & Neill, J.D. (1981). Dopamine levels in hypophysial stalk plasma and prolactin levels in peripheral plasma of the lactating rat: Effects of a simulated suckling stimulus. *Neuroendocrinology* **32**, 229-233.

de Greef, W.J., Voogt, J.L., Visser, T.J., Lamberts, S.W.J. & van der Schoot, P. (1987). Control of prolactin release induced by suckling. *Endocrinology* **121**, 316-322.

Demarest, K.T., Kckay, D.W., Riegle, G.D. & Moore, K.E. (1983). Biochemical indices of tuberoinfundibular dopaminergic neuronal activity during lactation: a lack of response to prolactin. *Neuroendocrinology* **36**, 130-137.

Di Scala-Guenot, D., Strosser, M.T. & Richard, P. (1987). Electrical stimulations of perifused magnocellular nuclei *in vitro* elicit Ca^{2+}-dependent, tetrodotoxin-insensitive release of oxytocin and vasopressin. *Neuroscience Letters* **76**, 209-214.

Dubois-Dauphin, M., Armstrong, W.E., Tribollet, E. & Dreifuss, J.J. (1985). Somatosensory systems and the milk-ejection reflex in the rat. II. The effects of lesions in the ventroposterior thalamic complex, dorsal columns and lateral cervical nucleus-dorsolateral fasciculus. *Neuroscience* **15**, 1131-1140.

Dyball, R.E.J. & Leng, G. (1986). Regulation of the milk ejection reflex in the rat. *Journal of Physiology* **380**, 239-256.

Erskine, M.S., Barfield, R.J. & Goldman, B.D. (1978). Intraspecific fighting during late pregnancy and lactation in rats and effects of litter removal. *Behavioral Biology* **23**, 206-218.

Freund-Mercier, M.J. & Richard, Ph. (1984). Electrophysiological evidence for facilitatory control of oxytocin neurones by oxytocin during suckling in the rat. *Journal of Physiology* **352**, 447-466.

Fukuoka, T., Negoro, H., Honda, K., Higuchi, T. & Nishida, E. (1984). Spinal pathway of the milk-ejection reflex in the rat. *Biology of Reproduction* **30**, 74-81.

Fuxe, K., Hökfelt, T. & Nilsson, O. (1969). Factors involved in the control of the activity of tuberoinfundibular dopamine neurons during pregnancy and lactation. *Neuroendocrinology* **5**, 257-270.

Greenfield, S.A. (1985). The significance of dendritic release of transmitter and protein in the substantia nigra. *Neurochemistry International* **6**, 887–901.

Grosvenor, C.E. & Mena, F. (1982). Regulation mechanisms for oxytocin and prolactin secretion during lactation. In: *Neuroendocrine Perspective*, Vol.1 (Müller, E.E. and MacLeod, R.M., eds.), Elsevier Biomedical Press, Amsterdam, pp. 69–110.

Hatton, G.I., Armstrong, W.E. & Gregory, W.A. (1978). Spontaneous and osmotically-stimulated activity in slices of rat hypothalamus. *Brain Research Bulletin* **3**, 497–508.

Higuchi, T., Bicknell, R.J. & Leng, G. (1991a). Reduced oxytocin release from the neural lobe of lactating rats is associated with reduced pituitary content and does not reflect reduced excitability of oxytocin neurons. *Journal of Neuroendocrinology* **3**, 297–302.

Higuchi, T., Honda, K., Fukuoka, T., Negoro, H., Hosono, Y. & Nishida, E. (1983). Pulsatile secretion of prolactin and oxytocin during nursing in the lactating rat. *Endocrinologia Japonica* **30**, 353–359.

Higuchi, T., Honda, K., Fukuoka, T., Negoro, H. & Wakabayashi, K. (1985). Release of oxytocin during suckling and parturition in the rat. *Journal of Endocrinology* **105**, 339–346.

Higuchi, T., Honda, K., Takano, S. & Negoro, H. (1988). Reduced oxytocin response to osmotic stimulus and immobilization stress in lactating rats. *Journal of Endocrinology* **116**, 225–230.

Higuchi, T., Honda, K., Takano, S. & Negoro, H. (1990). Stress-induced oxytocin release in the rat after lesion of the paraventricular nuclei; possible deficiency of corticotrophin-releasing factor. *Journal of Neuroendocrinology* **2**, 647–651.

Higuchi, T., Honda, K., Takano, S. & Negoro, H. (1992a). Abolishment of prolactin surge induced by ovarian steroid hormones in the lactating rat. *Neuroendocrinology*, in press.

Higuchi, H., Honda, K., Takano, S. & Negoro, H. (1992b). Estrogen fails to reduce tuberoinfundibular dopaminergic neuronal activity and to cause a prolaction surge in lactating, ovariectomized rats. *Brain Research*, in press.

Higuchi, T., Negoro, H. & Arita, J. (1989). Reduced responses of prolactin and catecholamine to stress in the lactating rat. *Journal of Endocrinology* **122**, 495–498.

Higuchi, T., Tadokoro, Y., Honda, K. & Negoro, H. (1986). Detailed analysis of blood oxytocin levels during suckling and parturition in the rat. *Journal of Endocrinology* **110**, 251–256.

Hyde, J.F., Murai I. & Ben-Jonathan, N. (1987). The rat posterior pituitary contains a potent prolactin-releasing factor: Studies with perifused anterior pituitary cells. *Endocrinology* **121**, 1531–1539.

Hyde, J.F., North, W.G. & Ben-Jonathan, N. (1989). The vasopressin-associated glyco-peptide is not a prolactin-releasing factor: Studies with lactating Brattleboro rats. *Endocrinology* **125**, 35–40.

Inenaga, K. & Yamashita, H. (1986). Excitation of neurones in the rat paraventricular nucleus *in vitro* by vasopressin and oxytocin. *Journal of Physiology* **370**, 165–180.

Jones, T.H., Brown, B.L. & Dobson, P.R.M. (1990). Paracrine control of anterior pituitary hormone secretion. *Journal of Endocrinology* **127**, 5–13.

Juss, T.S. & Wakerley, J.B. (1981). Mesencephalic areas controlling pulsatile oxytocin release in the suckled rat. *Journal of Endocrinology* **91**, 233–244.

Kaufman, S. (1981). Control of fluid intake in pregnant and lactating rats. *Journal of Physiology* **318**, 9–16.

Kawakami, M., Kimura, F. & Konno, T. (1973). Possible role of the medial basal prechiasmatic area in the release of LH and prolactin in rats. *Endocrinologia Japonica* **20**, 335–344.

Kawakami, M. & Sakuma, Y. (1976). Electrophysiological evidence for possible participation of periventricular neurons in anterior pituitary regulation. *Brain Research* **101**, 79–94.

Kiss, J.Z., Palkovits, M., Záborszky, L., Tribollet, E., Szabó, D. & Makara, G.B. (1983). Quantitative histological studies on the hypothalamic paraventricular nucleus in rats: I. Number of cells and synaptic boutons. *Brain Research* **262**, 217–224.

Lang, R.E., Heil, J.W.E., Ganten, D., Hermann, K., Unger, T. & Rasher, W. (1983). Oxytocin unlike vasopressin is a stress hormone in the rat. *Neuroendocrinology* **37**, 314–316.

Leng, G. (1988). Overture to synchrony and percussion. In: *Pulsatility in Neuroendocrine Systems* (Leng, G., ed.), CRC Press, Boca Raton, pp. 3–9.

Leng, G. & Shibuki, K. (1987). Extracellular potassium changes in the rat neurohypophysis during activation of the magnocellular neurosecretory system. *Journal of Physiology* **392**, 97–111.

Lenz, H.J., Raedler, A., Greten, H. & Brown, M.R. (1987). CRF initiates biological actions within the brain that are observed in response to stress. *American Journal of Physiology* **252**, R34–R39.

Leranth, C., Záborszky, L., Marton, J. & Palkovits, M. (1975). Quantitative studies on the supraoptic nucleus in the rat. *Experimental Brain Research* **22**, 509–523.

Lightman, S.L. & Young III, W.S. (1989). Lactation inhibits stress-mediated secretion of corticosterone and oxytocin and hypothalamic accumulation of corticotropin-releasing factor and enkephalin messenger ribonucleic acids. *Endocrinology* **124**, 2358–2364.

Lincoln, D.W., Hill, A. & Wakerley, J.B. (1973). The milk-ejection reflex of the rat: an intermittent function not abolished by surgical levels of anaesthesia. *Journal of Endocrinology* **57**, 459–476.

MacLeod, R.M. (1976). Regulation of prolactin secretion. In: *Frontiers in Neuroendocrinology* Vol. 4 (Martini, L. & Ganong, W.F., eds.), Raven Press, New York, pp. 169–194.

Moos, F., Ingram, C.D., Wakerley, J.B., Guerné, Y., Freund-Mercier, M.J. & Richard, Ph. (1991). Oxytocin in the bed nucleus of the stria terminalis and lateral septum facilitate bursting of hypothalamic neurons in suckled rats. *Journal of Neuroendocrinology* **3**, 163–171.

Moos, F., Poulain, D.A., Rodriguez, F., Guerné, Y., Vincent, J.-D. & Richard, Ph. (1989). Release of oxytocin within the supraoptic nucleus during milk ejection reflex in rats. *Experimental Brain Research* **76**, 593–602.

Moos, F. & Richard, Ph. (1989). Paraventricular and supraoptic bursting oxytocin cells in rat are locally regulated by oxytocin and functionally related. *Journal of Physiology* **408**, 1–18.

Mori, M., Vigh, S., Miyata, A., Yoshimura, T., Oka, S. & Arimura, A. (1990). Oxytocin is the major prolactin releasing factor in the posterior pituitary. *Endocrinology* **125**, 1009–1013.

Murai, I. & Ben-Jonathan, N. (1987). Posterior pituitary lobectomy abolishes the suckling-induced rise in prolactin (PRL): Evidence for a PRL-releasing factor in the posterior pituitary. *Endocrinology* **121**, 205–211.

Nagy, G.M. & Frawley, L.S. (1990). Suckling increases the proportions of mammotropes responsive to various prolactin-releasing stimuli. *Endocrinology* **127**, 2079–2084.

Nagy, G. & Halász, B. (1983). Time course of the litter removal-induced depletion in plasma prolactin levels of lactating rats. *Neuroendocrinology* **37**, 459–462.

Nagy, G., Mulchahey, J.J., Smyth, D.G. & Neill, J.D. (1988). The glycopeptide moiety of vasopressin-neurophysin precursor is neurohypophysial prolactin releasing factor. *Biochemical and Biophysical Research Communications* **151**, 524–529.

Negoro, H., Higuchi, T., Tadokoro, Y. & Honda, K. (1988). Osmoreceptor mechanism for oxytocin release in the rat. *Japanese Journal of Physiology* **38**, 19–31.

Negoro, H., Uchide, K., Honda, K. & Higuchi, T. (1985). Facilitatory effect of antidromic stimulation on milk ejection-related activation of oxytocin neurons during suckling in the rat. *Neuroscience Letters* **59**, 21–25.

Neill, J.D. (1988). Prolactin secretion and its control. In: *The Physiology of Reproduction* Vol. 1 (Knobil, E. & Neill, J.D., eds.), Raven Press, New York, pp. 1379–1390.

Nikolics, K., Mason, A.J., Szönyi, E., Ramachandran, J. & Seeburg, P.H. (1985). A prolactin-inhibiting factor within the precursor for human gonadotropin-releasing hormone. *Nature* **316**, 511–517.

Patel, H., Chowdrey, H.S. & Lightman, S.L. (1991). Lactation abolishes corticotropin-releasing factor-induced oxytocin secretion in the conscious rat. *Endocrinology* **128**, 725–727.

Plotsky, P.M. & Neill, J.D. (1982a). The decrease in hypothalamic dopamine secretion induced by suckling: comparison of voltammetric and radioisotopic methods of measurements. *Endocrinology* **110**, 691–696.

Plotsky, P.M. & Neill, J.D. (1982b). Interactions of dopamine and thyrotropin-releasing hormone in the regulation of prolactin release in lactating rats. *Endocrinology* **111**, 168–173.

Poulain, D.A. & Wakerley, J.B. (1986). Afferent projections from the mammary glands to the spinal cord in the lactating rat II. Electrophysiological responses of spinal neurons during stimulation of the nipples, including suckling. *Neuroscience* **19**, 511–521.

Pow, D.W. & Morris, J.F. (1989). Dendrites of hypothalamic magnocellular neurons release neurohypophysial peptides by exocytosis. *Neuroscience* **32**, 435–439.

Randle, J.C.R., Bourque, C.W. & Reneud, L.P. (1986). α_1-adrenergic receptor activation depolarizes rat supraoptic neurosecretory neurons *in vitro*. *American Journal of Physiology* **251**, R569–R574.

Rondeel, J.M.M., de Greef, W.J., Visser, T.J. & Voogt, J.L. (1988). Effect of suckling on the in vivo release of thyrotropin releasing hormone, dopamine and adrenaline in the lactating rat. *Neuroendocrinology* **48**, 93–96.

Samson, W.K., Lumpkin, M.D. & McCann, S.M. (1986). Evidence for a physiological role for oxytocin in the control of prolactin secretion. *Endocrinology* **119**, 554–560.

Saphier, D. & Feldman, S. (1985). Electrophysiological evidence for neural connections between the paraventricular nucleus and neurons of the supraoptic nucleus in the rat. *Experimental Neurology* **89**, 289–294.

Schally, A.V., Redding, T.W., Arimura, A., Dupont, A. & Linthicum, G.L. (1977). Isolation of gamma-amino butyric acid from pig hypothalami and demonstration of its prolactin release-inhibiting (PIF) activity *in vivo* and *in vitro*. *Endocrinology* **100**, 681–691.

Stern, J.M. & Voogt, J.L. (1973/74). Comparison of plasma corticosterone and prolactin levels in cycling and lactating rats. *Neuroendocrinology* **13**, 173–181.

Takano, S., Negoro, H., Honda, K. & Higuchi, T. (1990a). Lesion study of milk ejection pathway in the mid-hypothalamus of rat. *Japanese Journal of Physiology* **40**

(Supplement), S100.

Takano, S., Negoro, H., Honda, K. & Higuchi, T. (1990b). Electrophysiological evidence for neural connections between the supraoptic nuclei. *Neuroscience Letters* **111**, 122–126.

Takano, S., Negoro, H., Honda, K. & Higuchi, T. (1991). Another milk ejection-related burst of the neurone in the hypothalamus. *Neuroscience Research* Supplement **14**, S101.

Terkel, J., Blake, C.A. & Sawyer, C.H. (1972). Serum prolactin levels after suckling or exposure to ether. *Endocrinology* **91**, 49–53.

Theodosis, D.T. (1985). Oxytocin-immunoreactive terminals synapse on oxytocin neurones in the supraoptic nucleus. *Nature* **313**, 682–684.

Theodosis, D.T. & Poulain, D.A. (1984). Evidence for structural plasticity in the supraoptic nucleus of the rat hypothalamus; in relation to gestation and lactation. *Neuroscience* **11**, 183–193.

Thoman, E.B., Conner, R.L. & Levine, S. (1970). Lactation suppresses adrenal corticosteroid activity and aggressiveness in rats. *Journal of Comparative and Physiological Psychology* **70**, 364–369.

Tindal, J.S. & Knaggs, G.S. (1977). Pathways in the forebrain of the rat concerned with the release of prolactin. *Brain Research* **119**, 211–221.

Tindal, J.S. & Knaggs, G.S. (1981). Determination of the detailed hypothalamic route of the milk-ejection reflex in the guinea-pig. *Journal of Endocrinology* **50**, 135–152.

Tribollet, E., Barberis, C., Jard, S., Dubois-Dauphin, M. & Dreifuss, J.J. (1988). Localization and pharmacological characterization of high affinity binding sites for vasopressin and oxytocin in the rat brain by light microscopic autoradiography. *Brain Research* **442**, 105–118.

Verbalis, J.G., McCann, M.J., McHale, C.M. & Stricker, E.M. (1986a). Oxytocin secretion in response to cholecystokinin and food: differentiation of nausea from satiety. *Science* **232**, 1417–1419.

Verbalis, J.G., McHale, C.M., Gardiner, T.W. & Stricker, E.M. (1986b). Oxytocin and vasopressin secretion in response to stimuli producing learned taste aversions in rats. *Behavioral Neuroscience* **100**, 466–475.

Wakerley, J.B., Clarke, G. & Summerlee, A.J.S. (1988). Milk ejection and its control. In: *The Physiology of Reproduction* (Knobil, E. and Neill, J.D., eds.), Raven Press, New York, pp. 2283–2321.

Wakerley, J.B., Juss, T.S., Farrington, R. & Ingram, C.D. (1990). Role of the paraventricular nucleus in controlling the frequency of milk ejection and the facilitatory effect of centrally administered oxytocin in the suckled rat. *Journal of Endocrinology* **125**, 467–475.

Wakerley, J.B. & Lincoln, D.W. (1973). The milk-ejection reflex of the rat: a 20- to 40-fold acceleration in the firing of paraventricular neurones during oxytocin release. *Journal of Endocrinology* **57**, 477–493.

Wakerley, J.B. & Negoro, H. (1990). Evidence for altered sensitivity of supraoptic neurones to phenylephrine in lactating versus non-lactating rats. *Neuroendocrinology* **52** (supplement 1), 118.

Yokoyama, A. & Ôta, K. (1959). Effect of oxytocin replacement on lactation in rats bearing hypothalamic lesions. *Endocrinologia Japonica* **6**, 268–276.

Index